Mihail Shojhedbrod
Irina Shojhedbrod

Microgravity electric separator

Mihail Shojhedbrod
Irina Shojhedbrod

Microgravity electric separator

Development prospective life support and power supply systems for the International Space Station space station

ScienciaScripts

Imprint

Cover image: www.ingimage.com

This book is a translation from the original published under ISBN 978-620-8-43915-6.

Publisher:
Sciencia Scripts
is a trademark of
Dodo Books Indian Ocean Ltd. and OmniScriptum S.R.L publishing group

120 High Road, East Finchley, London, N2 9ED, United Kingdom
Str. Armeneasca 28/1, office 1, Chisinau MD-2012, Republic of Moldova, Europe
Managing Directors: Ieva Konstantinova, Victoria Ursu
info@omniscriptum.com

Printed at: see last page
ISBN: 978-620-6-17190-4

Microgravity electric separator

Michael Shoikhedbrod
Irina Shoikhedbrod

2025

Dedicated

Our family:

Elena and Aiden, Igor, Julia and Sarochka, Ariel and Richelle,

CONTENTS.

INTRODUCTION

Tasks of life support and power supply systems of the International Space Station

Oxygen and water are vital to the smooth operation of the crews of the International Space Station and interplanetary spacecraft during long missions.

Replenishing water, oxygen and microalgae biomass by delivering them to the International Space Station is not economically viable.

Therefore, there is a problem of using and developing new systems that utilize regeneration of oxygen, water from astronauts' life products, purification of liquid fuel and technical water from harmful solid inclusions and microbes, cultivation of rapidly multiplying microalgae on the International Space Station itself or on an interplanetary spacecraft [1].

The use of microalgae biomass solves the problem of utilization of carbon dioxide formed in the process of astronauts' life activity, replenishment of the spacecraft atmosphere with oxygen and replenishment of astronauts' nutrition [2].

Recovery of water from urine in the life support system of automatic spacecraft using electric field effects on urine under conditions

Water reclamation is the re-production of water, which is vital to the normal functioning of astronauts on the International Space Station (ISS)

Initially, water was originally delivered to the ISS from Earth, which was too expensive and therefore weight-limited. Therefore, the regeneration process itself reduces the cost of space travel and also makes the ISS system less dependent on ground services.

Several methods of water regeneration are currently used on the ISS:

- Condensation of moisture from the air;

- Treatment of used water;

- Urine and solid waste recycling.

The ISS has special equipment that condenses moisture from the air.

Water purification plays a major role in regeneration. Any water on the ISS is collected in purification systems - leftovers from cooking, dirty water from washing and even astronauts' sweat,

All this water is collected in a special distiller, which visually resembles a barrel. When purifying the water by rotating the distiller, artificial gravity is created and the dirty water is run through filters. The result is clean drinking water.

In the last step, iodine is added to this water. This chemical helps to prevent the growth of germs and bacteria and is also an essential element for the health of astronauts.

Due to the complexity of storing liquid wastes of astronauts' activity on the ISS and a sufficiently large mass of water for technical purposes, for example, for the operation of catalysts and obtaining oxygen for the air breathed by astronauts, the regeneration of water from urine is of great interest.

There are several methods of water recovery from urine. A. Paknat in [3] presented a study of water recovery by electrolytic purification of urine.

Experiments were performed leading to the calculated parameters of the urine electrolysis module.

These experiments showed that intermittent electrode operation or intermittent exposure of urine to continuously operating electrodes gave the best current performance.

Using a 3 min to 6 min off-cycle at 1.2 A/cm2 and 0.05 A/ml or approximately 85% current efficiency was obtained.

Two module designs using platinized platinum electrodes have been proposed.

The assumed electrode arrangement places a platinum wire anode concentrically inside a perforated cylindrical Pt cathode with an annular gap of 0.2 cm.

The preferred module design utilizes a centrifugal pump to circulate urine within the electrolysis module.

The module consists of an electrolysis chamber, a reaction chamber, a gas-liquid phase separator and a centrifugal pump.

When this module is operating at maximum efficiency, electrolysis alone will consume 2,767 kilowatt-hours per six-liter batch of average urine for 7.73 hours.

That's compared to 1,674 kilowatt-hours for a six-hour process, which is theoretically possible.

Analysis of gases emitted during electrolysis of real and "complex" synthetic urine showed the CO_2/N_2 ratio expected for urea oxidation.

They also showed that real urine undergoes electrolytic oxidation less efficiently than synthetic urine.

Analysis of coliform bacteria in raw urine and the same urine after electrolysis showed that both were virtually free of coliform bacteria.

D.Brown et al. in [4] demonstrated the extraction of water from urine using membrane electrodialysis.

A laboratory pilot program was conducted that clearly indicated the feasibility of using membrane electrodialysis to extract potable water from urine.

Since electrodialysis removes electrolytes from urine, pretreatment methods were investigated to remove non-electrolytes.

These methods were: adsorption on charcoal, chemical oxidation and enzymatic hydrolysis of urea to ammonium carbonate.

A model of a urine recycling system consisting of two main processes, adsorption on carbon and electrodialysis, was constructed.

The system weighs 15 pounds, is smaller than a single Fit and has a peak power consumption of 45 watts for 8 hours of operation to extract 3,000 ml of drinking water from urine each day.

Daily energy consumption was 310Wh, including 160Wh for coal regeneration.

Water of excellent quality was extracted in 92% yield. D. Putnam and others in [5] presented the extraction of water from urine by chromium ion treatment and distillation.

An improved method for extracting drinking water from urine by distillation was proposed, in which the urine was pretreated prior to distillation with a urine-soluble hexavalent chromium compound alone or in combination with a urine-soluble sulfate compound.

P. Budininkas in [6] showed the four-man rated dual catalyst system for the recovery of water from urine.

The catalytic system was integrated with a 4-person wick urine vaporizer.

During operation, urine vapor produced by the wick evaporator was treated in a catalytic system to remove ammonia and volatile hydrocarbons, and water was extracted by condensation in a water-cooled condenser.

The system was fully automatic and did not require any manual adjustments except for periodic delivery of urine and removal of extracted water.

Although the system was designed to process 0.325 kg of urine per hour, this rate could only be achieved with a fresh wick and then gradually decreased as the wick became saturated with urine solids.

The mean urine handling rates achieved during each of the three endurance trials were 0.137, 0.217, and 0.235 kg/hour.

The quality of the extracted water met drinking water standards except for generally low pH.

Currently, the Russian segment of the ISS uses a modernized WRS-UM (Water Recovery System - Urine Modernized) unit [7], which operates according to the following principle: the device contains a number of vacuum stages-containers that simultaneously rotate.

The liquid under the action of centrifugal forces is thrown to the periphery, and the liquid-gas mixture is separated.

The dried steam from the first stage enters the second stage for more reliable cleaning of fine droplets and so on four times. The result is a water recovery of up to 90%.

The American UPA (Urine Processor Assembly) system [8] used on the American segment of the ISS is based on the process of vacuum distillation in a centrifugal distiller with a rotating cylindrical surface, on one side of which the process of water evaporation from urine takes place, and on the other side - vapor condensation.

The temperature difference for condensation heat transfer to evaporation is created by steam compression in a steam compressor built into the distiller. Thus, it should be noted that modern urine water extraction systems are based on a relatively energy-consuming distillation process, so one of the main design objectives is to reduce the energy consumption of distillation.

Cleaning of fuel fluid in the power supply system and technical water in the life support system of the International Space Station from contaminants and microbial pollution

The purification of liquid propellants from contaminants in the rocket engine fuel tank and fluids in the life support system of automatic spacecraft is a very important and also critical task for automatic spacecraft[9, 10].

No less important is the purification of liquid propellant from microorganisms actively multiplying in the tanks of liquid propellant spacecraft engines or in the life support system fluids of automatic spacecraft, causing fouling, failures and corrosion in the life support system storage tanks, pipelines and engines [11, 12].

Water reclamation is the re-production of water, which is vital to the normal functioning of astronauts on the International Space Station (ISS)[10].

Initially, water to the ISS was supplied from Earth, which was too expensive and therefore weight-limited. Therefore, the regeneration process itself reduces the cost of space travel and the ISS system becomes less dependent on ground services.

Existing methods and devices for purifying fuel fluid from contaminants or microbes, recovering water from the International Space Station's life support system service water[9-12] are expensive and uneconomical.

Concentrate of fast-growing microalgae production for conversion of carbon dioxide into oxygen and food biomass to feed the crew of the International Space Station during a long-duration mission.

It is known that the use of microalgae solves the problem of utilization of carbon dioxide formed in the process of life activity of astronauts and replenishment of oxygen in the atmosphere of a spacecraft. There is evidence that among the mass of microalgae used for research as oxygen products for space flights, chlorella and spirulina are suitable.

In the field of the International Space Station (ISS) Life Support System (LSS), research is conducted mainly for the ongoing support of manned missions on the Russian segment of the international station.

Spirulina is mainly used as a biological supplement in a mixture with various components. In the experimental complex "Bios-3" [13] a number of experiments in hermetic chambers were conducted, such as 45-day experiment with two-link human-microalgae system, 5-month experiment with two-link human-high plants system, the duration of the experiment was 1.5 years of three-link human-microalgae-high plants system.

As a result of medico-technical experiments it is shown that the biological life support system based on unicellular microalgae and higher plants is fully loaded with oxygen, water and plant food.

No adverse environmental effects on the body of the testers were detected.

Currently, experiments are being conducted on the ISS using the space photobioreactor shown in Figure 1 [14] to determine the possibility of using microalgae to convert carbon dioxide emitted by astronauts into oxygen and food biomass by photosynthesis.

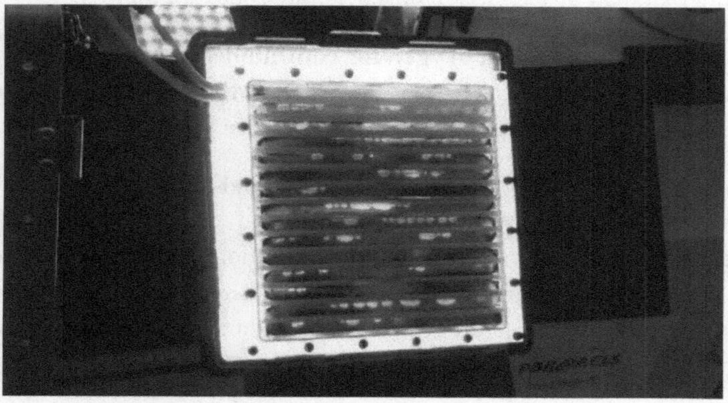

Figure 1: Space photobioreactor

The microalgae-based photobioreactor is an important step towards a closed-loop life-support system that would eliminate the need to ship supplies from Earth. This is especially important for future long-term missions to the Moon or Mars, which require more materials than a spacecraft can accommodate.

The photobioreactor works in conjunction with the physicochemical air recirculation system or Advanced Closed Loop System (ACLS), which was delivered to the ISS in 2018. ACLS gets methane and water from the carbon dioxide in the cabin.
The microalgae in the photobioreactor use the remaining carbon dioxide to generate oxygen, creating a hybrid solution formally known as PBR@ACLS.

However, only cultivation of microalgae under microgravity conditions is not enough, due to the fact that a concentrate of live microalgae is required for astronauts' nutrition, and in order to maintain microalgae cultivation under long flight conditions, it is necessary to add a part of the obtained concentrate and separated nutrient medium to the cultivator (photobioreactor) for microalgae cultivation in a closed cycle.

In this regard, there is a need to develop new methods and devices to produce concentrate of rapidly multiplying microalgae for processing the released carbon dioxide into oxygen and food biomass to feed the crew of the International Space Station during a long space flight.

Obtaining hydrogen and oxygen for continuous charging of fuel cells to supply the power system of the International Space Station in microgravity conditions.

Any flight control of a spacecraft (SC) depends on the availability of electricity on board. All its instruments and equipment (propulsion, controls, communication systems, instrumentation, heating, etc.) on board the spacecraft function with electricity. The electrical power supply system (EPS) of a spacecraft is the most important system ensuring its life support.

The spacecraft power supply system is based on a system that generates direct electric current. This is due to the fact that most of the sources that can be used on board are DC sources. At present, the main one is a source that converts any energy (chemical, light, nuclear) into electrical energy, and which should ensure the astronauts' work during the entire flight.
Today, three sources that convert chemical energy directly into electrical energy with rather high efficiency (about 70%) have found their practical application: galvanic cells, batteries, and fuel cells taken from the Earth [15].

The most developed are fuel cells, which utilize hydrogen and oxygen as "fuel and oxygen" [16, 17].

The fuel cell (Figure 2) consists of two electrodes separated by electrolyte and systems for supplying fuel to one electrode and oxidizer to the other, as well as systems for removing reaction products.

Figure 2 illustrates a fuel cell with an acid electrolyte in which hydrogen is supplied through a hollow anode and enters the electrolyte through very small pores in the electrode material. This decomposes the hydrogen molecule into atoms, which are converted into positively charged ions by chemisorption, giving up one electron each.

This process can be described by the following equations:

$$H_2 \leftrightarrow 2H \quad (1)$$

$$2H = 2H^+ + 2e^- \quad (2)$$

Hydrogen ions diffuse through the electrolyte to the positive element.

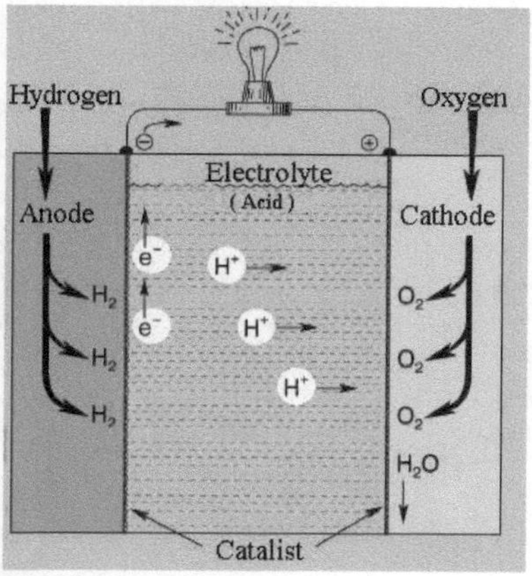

Figure 2: Fuel cell with acid electrolyte

The oxygen supplied to the cathode passes into the electrolyte and also reacts on the electrode surface with the participation of the catalyst. The oxygen combines with hydrogen ions and electrons from the external circuit to form water:

$$1/2O_2 + 2H^+ + 2e^- \rightarrow H_{(2)}O \quad (3)$$

In fuel cells with an alkaline electrolyte (usually concentrated sodium or potassium hydroxides) (Figure 3), similar chemical reactions occur. Hydrogen passes through the anode and, with the help of a catalyst, reacts with the ($OH^{-)}$ ions present in the hydroxyl electrolyte to form water and electrons:

$$H_2 + 2OH^- \rightarrow 2H_2O + 2e^- \quad (4)$$

At the cathode, oxygen reacts with water in the electrolyte and electrons from an external circuit.

In successive reaction steps, hydroxyl ions (as well as perhydroxyl O_2H^-) are formed.

The reaction at the cathode can be written as:

$$1/2O_2 \quad + \quad H_{(2)}O \quad + \quad 2e^{(-)} \quad \rightarrow \quad 2OH^{(-)})$$

(5)

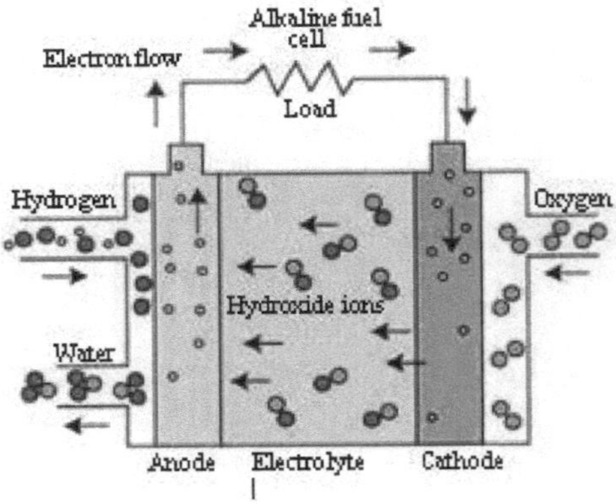

Figure 3: Alkaline fuel cell

The flow of electrons and ions maintains the balance of charge and matter in the electrolyte. The water produced by the reaction partially dilutes the electrolyte.

In a fuel cell, some of the energy of the chemical reaction is converted to heat, and the resulting flow of electrons in the external circuit is the direct current used to power the onboard instruments and equipment of the spacecraft. Most reactions in fuel cells provide an electromotive force (EMF) of about 1 V.

Opening the circuit or stopping the movement of ions stops the fuel cell from operating.

The process that occurs in a hydrogen-oxygen fuel cell is essentially the reverse of the process of electrolysis, in which water dissociates under the action of an electric current, i.e., when voltage is applied to the electrodes of a hydrogen-oxygen fuel cell; water can decompose into hydrogen and oxygen, which can collect at the electrodes.

If you stop charging the cell and connect a load to it, such a regenerative fuel cell will immediately start operating in normal mode.

Theoretically, the size of the fuel cell can be as large as desired.

In practice, however, multiple cells are combined into small modules or batteries that are connected either in series or parallel.

Two important advantages of fuel cells over electric batteries:

- Functioning until the fuel (combustible and oxidizer - hydrogen and oxygen) supplied from an external source is fully utilized;

- The fuel cell does not need to be recharged during operation.

Under spaceflight conditions, hydrogen and oxygen are continuously supplied to the fuel cells from tanks refueled on Earth with liquid hydrogen and oxygen.

The Apollo spacecraft service module energy subsystem consists of two hydrogen tanks, two oxygen tanks, and three fuel cells (Figure 4).

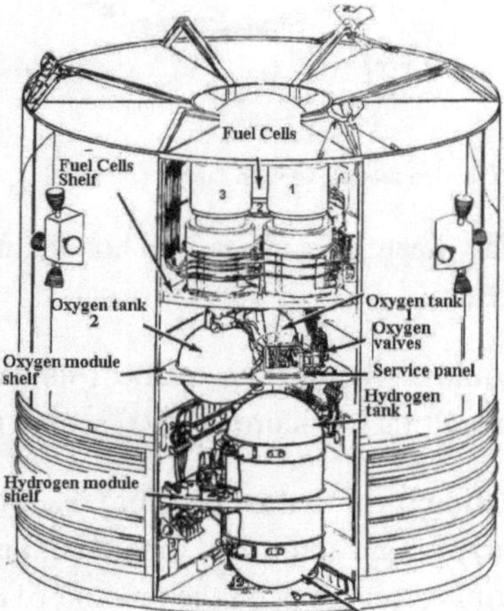

Figure 4: Energy subsystem of the spacecraft service module

Fuel cells, by consuming hydrogen and oxygen, generate electricity and water, which is used by the crew for drinking and cooling equipment.

The Apollo Service Module oxygen tank is shown in Figure 5.

Liquid oxygen is stored in the tank as a supercritical fluid and therefore exhibits both liquid and gas properties.

As the gas expands, the temperature decreases. Thermal insulation is so good that liquid oxygen loses its supercritical properties when cooled simply by expansion at normal fuel cell flow rates.

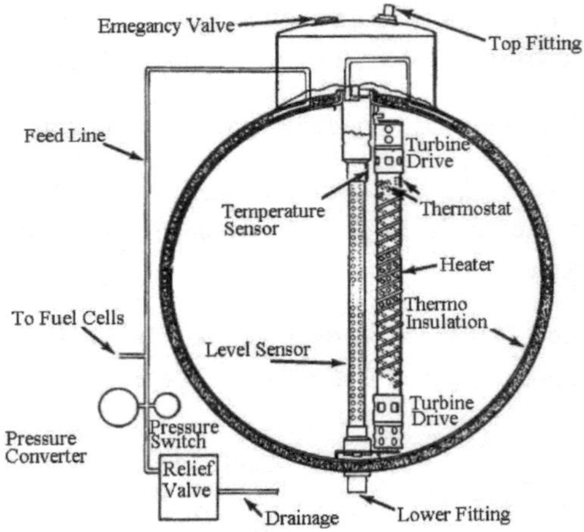

Figure 5: Apollo spacecraft service module oxygen tank

Therefore, a special heater is installed to maintain the required temperature and pressure.

In weightlessness, liquid oxygen in supercritical state stratifies into liquid and gaseous layers, resulting in incorrect level sensor readings.

A special oxygen mixing turbine was therefore installed, which added to the crew's need to carry out an additional oxygen mixing procedure in the tanks in order to obtain correct data on the amount of oxygen on board.

All this leads to expensive delivery of oxygen tanks on board the spacecraft, greatly complicates the time-limited and unsafe supply of hydrogen and oxygen to the fuel cells (the accident during the Apollo 13 flight, [18]).

Advanced life support and electrical power systems for long duration missions on the International Space Station utilizing the Universal Electrical Separator System

The universal electroseparator developed by the author made it possible to create a new life support system (LSS) for the International Space Station (ISS) or an interplanetary spacecraft, which includes a whole complex of universal electroseparators providing fuel liquid purification and drinking water production by purifying dirty technical water from solid harmful impurities or microbes, as well as producing drinking water from astronauts' urine in microgravity in quantities sufficient for astronauts to consume this water.

The life support system (LSS) of the International Space Station (ISS) or interplanetary spacecraft under development has successfully passed flight tests aboard the IL-76K flying laboratory and demonstrated its effective operation in short-term weightlessness.

In this book, the authors reveal the physical principles of operation of the universal electroseparator in weightlessness and its design; present how the 24-hour purification of water from urine in the life support system of the International Space Station in microgravity using the universal electroseparator is carried out; show how the 24-hour purification of fuel fluid in the power supply system and technical water in the life support system of the International Space Station from contaminants and micropollutants is carried out.

CHAPTER I

PHYSICAL PRINCIPLES OF OPERATION OF THE UNIVERSAL ELECTRIC SEPARATOR IN WEIGHTLESSNESS AND ITS DESIGN

1.1 Physical principles of operation of universal electric separator in microgravity conditions

Water in the universal electric separator dissociates into ions under microgravity conditions, just like on Earth:

$$H_2O\ H\leftrightarrow^{+}+\ OH^{-} \qquad (1.1.1)$$

When a direct current is passed through water, one hydrogen ion H^+ approaches the cathode, robbing it of an electron, and four OH^- ions approach the anode, which give up four electrons.

As a result, the neutral hydrogen atoms H formed at the cathode in the free state become unstable and unite into pairs of the two-atom hydrogen molecule $H2$.

Two water molecules are formed at the anode, and the resulting neutral oxygen atoms O in the free state become unstable and pair up to form the two-atom oxygen molecule O_2.

As a result, hydrogen bubbles will bounce off at the cathode and oxygen bubbles will bounce off at the anode:

$$H\ e^{+} + \ = \ \mathrm{H};\quad H\,H + \ = \ H_{(2)}\uparrow \qquad (1.1.2)$$

$$4OH^{-} - 4e = \ 2H_2O + \ O_{(2)}\uparrow \qquad (1.1.3)$$

The physical mechanism for the formation of negatively charged hydrogen bubbles at the cathode and positively charged oxygen bubbles is as follows [19].

A hydrogen molecule in the form of a bubble formed at the cathode is immersed in a cloud of positively charged hydrogen ions H^+ , which induces a negative charge on the outer surface of the bubble, while the inner surface of the bubble will be positively charged.

The negative potential of the cathode will detach the negative charges of the outer surface of the bubble, pushing them to the top of the bubble, and attract the positive charges of the inner surface of the bubble, collecting them in the region near the cathode, as shown in Figure 6[20].

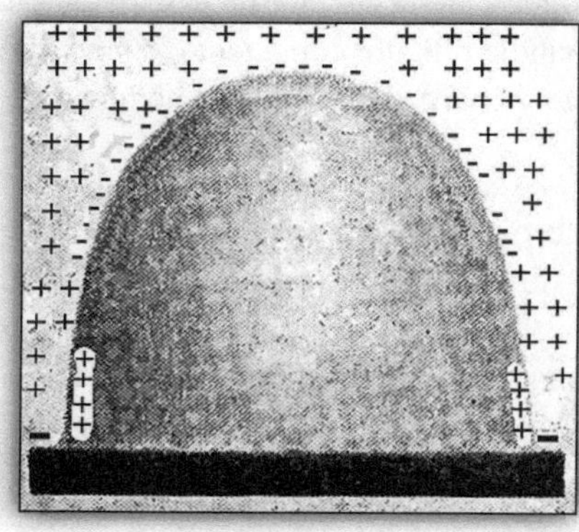

Figure 6: Photograph of a hydrogen bubble with contact angle θ=97° at the cathode and the catholyte solution (living water) formed around it (7.5X magnification) [20].

In Figure 6, the sign "+" denotes hydrogen cations H+, positive charges of the inner surface of the bubble on the cathode; the sign "-" denotes negative charge - charge of the cathode, negative charges of the outer surface of the bubble.

At the moment of bubble detachment, the positively charged near-cathode part of the bubble remains on the cathode, and the detached hydrogen bubble with negatively charged surface moves upward to the free water surface.

Intensive formation of negatively charged microdispersed electrolysis hydrogen bubbles, their repulsion from the cathode and rise to the free surface of water saturates water with gaseous hydrogen bubbles, forming negatively charged hydrogen water - catholyte solution (living water).

The neutral oxygen molecule formed at the anode in the form of a bubble is immersed in a cloud of negatively charged ***OH*** hydroxyls⁻ , which induces a positive charge on the outer surface of the bubble, while the inner surface of the bubble will be negatively charged. The positive potential of the anode, through electrostatic repulsion force, will displace the positive charges on the outer surface of the bubble towards the top of the bubble, and attract the negative charges on the inner surface of the bubble, collecting them in the anode region.

At the moment of bubble detachment, the negatively charged near-anode region part of the bubble remains on the anode, and the detached oxygen bubble with positively charged surface moves upwards to the free water surface.

In this case, the formation of positively charged electrolytic oxygen bubbles, their repulsion from the anode and rise to the free surface of the water saturates the water with gaseous oxygen bubbles, forming positively charged oxygen water - anolyte solution (dead water).

Thus, the bubbles formed at the electrode are always charged equally with the electrode, i.e., hydrogen bubbles take a negative charge and oxygen bubbles take a positive charge.

The conducted experiments [21] confirmed this conclusion.

The bubbles formed on the electrode during electrolysis have different sizes.

The reason for this difference in the size of the bubbles formed on the electrode is down to the different electrostatic interaction of the charged bubbles with the electrode.

The following forces act on the bubble formed during electrolysis and fixed on the electrode surface in the liquid: the bubble surface tension force $F\sigma$, by means of which the bubble is held on the electrode surface, acts on the perimeter of the bubble attachment and is equal to

$F\sigma = \pi a \sigma Sin(\theta)$, where: a = diameter of the circle where the bubble is fixed on the electrode surface; θ = contact angle at the metal-solution-gas interface; σ = surface tension at the solution-gas interface.

F_E-electrostatic force of repulsion of a hydrogen bubble from the cathode and an oxygen bubble from the anode.

F_{E1} - electrostatic force of attraction of the cloud of positive hydrogen ions to the negatively charged outer layer of a hydrogen bubble and the cloud of negative hydroxyl anions to the positively charged outer layer of an oxygen bubble, facilitating the detachment of bubbles from the electrodes

The force of hydrostatic lifting of a bubble by liquid according to Archimedes' law in microgravity is zero.

Depending on the sign of the charge of the electrode and the bubble, the bubble is either attracted by the electrode (electrostatic attraction force) or detached from it (electrostatic detachment force) and, accordingly, reaches a higher or lower value before detaching from the electrode.

Then the resulting force acting on the bubble at the points of its contact with the electrode surface can be represented as follows:

$$F\sigma = F(E)1 + F(E) \quad (1.1.4)$$

For microgravity, the contribution of F_{E1} and F_E to the bubble detachment from the electrode is significant.

Thus, the higher the electrode potential and the higher the potential of the cloud of positive cations or negative anions, the smaller the size of the bubble, because the higher the electrode potential, the higher the electrostatic detachment force acting between the electrode and the bubble formed on the surface of the electrode of the same charge, the detachment of the bubble from the electrode surface, and the faster the bubble detaches.

Thus, the size of the negatively charged hydrogen bubbles can be controlled by increasing the electric potential, making the bubbles as small as the process requires.

Studies have shown that the use of negatively charged hydrogen bubbles with calculated dispersity solves the main problem of any technological process associated with the separation of nutrient medium or concentration of solid microparticles or cells of microscopic organisms from the nutrient medium: the creation of strong complexes hydrogen bubbles + solid microparticle, microalgae cell.

The physical mechanism for the creation of this robust complex is as follows.

After formation and detachment from the surface of the cathode, negatively charged hydrogen bubbles, moving in the aqueous solution in the horizontal direction to the positively charged anode due to the force of electrostatic attraction in weightlessness, meet on their way a solid microparticle or microalgae cell, the size of which is much larger, and induce a positive charge on the surface of the particle of solid material or are attracted to the positive outer layer of the phospholipid bilayer of the microalgae cell membrane.

Then, as a result of the attraction of opposite charges, the bubbles adhere to the microparticle, forming a strong contact with it (Figure 7).

As a result of the attraction of opposite charges, the vesicles adhere to the positive outer layer of the phospholipid bilayer of the microalgae cell membrane (Figure 8).

The strength of the contact is due to the action of two forces: surface tension force $\boldsymbol{F\sigma = \pi a\sigma Sin(\theta)}$ and electrostatic attraction force $\boldsymbol{F_E}$, which together act to firmly fix negatively charged hydrogen bubbles on the surface of a solid microparticle or on the outer layer of the phospholipid bilayer of the microalgae cell membrane.

Hydrogen bubble microdispersion allows complete coverage of the entire surface of a solid microparticle or the outer layer of the phospholipid bilayer of the microalgae cell membrane if the size of the particle or microalgae cell is much larger than the size of the hydrogen bubble (Figure 9).

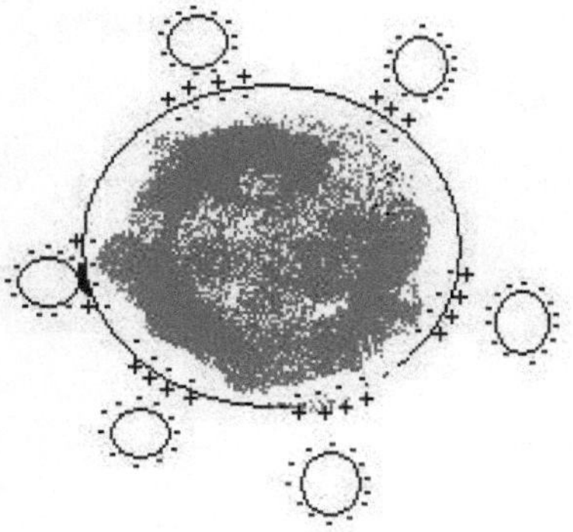

Figure 7: A negatively charged hydrogen bubble induces a positive charge on the surface of a solid microparticle and attaches to it under microgravity conditions

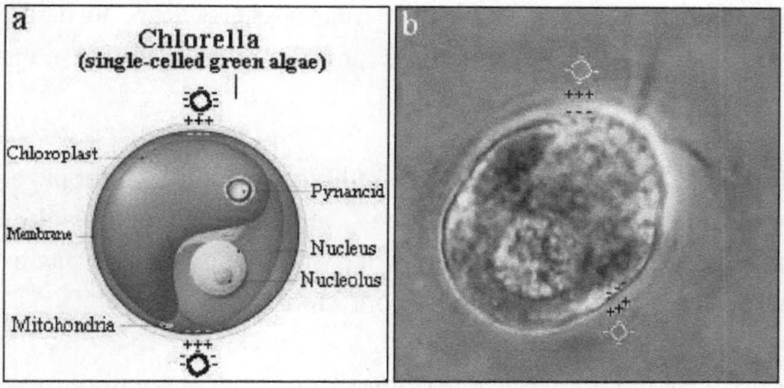

Figure 8: Complex formation: negatively charged hydrogen bubble + Chlorella cell (a) and Dunaliella cell (b) in Microgravity

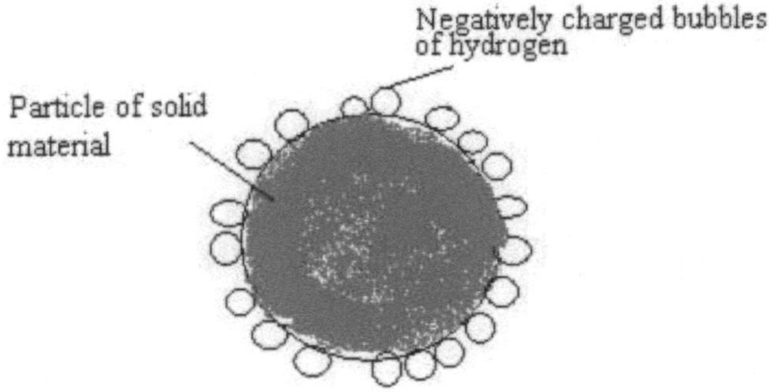

Figure 9: Solid microparticle coated with hydrogen bubbles if the microparticle size is large

As a result, negatively charged complexes of hydrogen bubbles + solid microparticle or hydrogen bubbles + microalgae cell are formed due to shielding of the solid microparticle or microalgae cell by negatively charged hydrogen bubbles.

The formed complexes hydrogen bubbles + solid microparticle or hydrogen bubbles + microalgae cell move in aqueous solution in horizontal direction to positively charged anode due to electrostatic attraction force under microgravity conditions.

If the size of a solid microparticle or microalgae cell is commensurate with the size of a negatively charged hydrogen bubble, the bubble attached to the surface of the particle or cell moves horizontally in the aqueous solution toward the positively charged anode for the same reason.

The fact of formation of negatively charged microdispersed hydrogen bubbles in the process of electrolysis, the movement of formed complexes towards the positively charged anode in aqueous solution under weightlessness conditions were experimentally proved during flight tests on board the flying laboratory (FL) IL-76K [21].
Figure 10 shows the formation of negatively charged microdispersed hydrogen bubbles and their rise to the indicated mark as a result of

electrostatic attraction of the anode and their retention at the indicated mark by vibration under short-term weightlessness conditions.

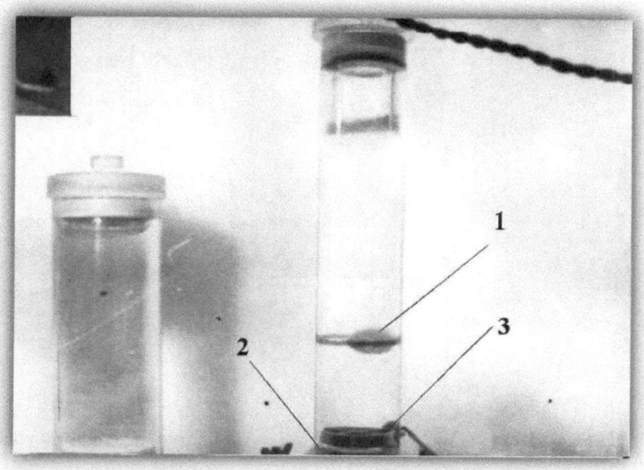

Figure 10: Photograph of the formation of negatively charged hydrogen bubbles during electrolysis and their rise to the indicated mark as a result of electrostatic attraction of a positively charged anode, keeping them at the indicated mark by vibration under short-term weightlessness conditions
1 - Localized negatively charged hydrogen bubbles
2 - Sub-electrode (cathode)
3 - Upper electrode (anode)

The lack of Archimedes' force creates a problem in delivering the complexes to the receiving manifold.

The problem was solved by using a universal electric separator and receiver collector capsules of conical shape.

Under microgravity conditions, a top-down force equal to $F = \sigma L$ acts on the aqueous solution contained in the cone-shaped chamber of a universal separator.

From bottom to top, a force equal to $F1 = \sigma L1$ acts on the water mixture.

Since **L>L1**, ***F>F1*** and hence the resultant ***force Fr = F-F1*** will be directed downward and the resulting water mixture will flow gently downward as shown in Figure 11.

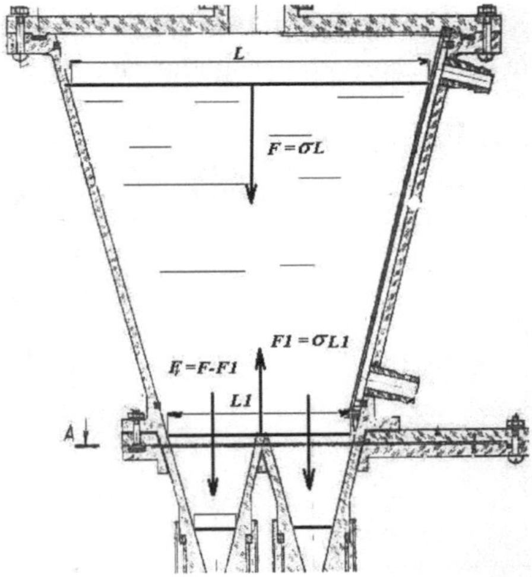

Figure 11: Movement of the urine mixture in the separator under weightless conditions

Thus, under microgravity conditions, one of the receiving collector capsules (anode zone) will be filled with separated complexes of negatively charged microdispersed hydrogen bubbles + solid microparticle or microalgae cell, and the other capsule will be filled with extracted purified water.

1.2 Design of the universal electric separator

Figures 12, 13 show the general shape of the universal electric separator

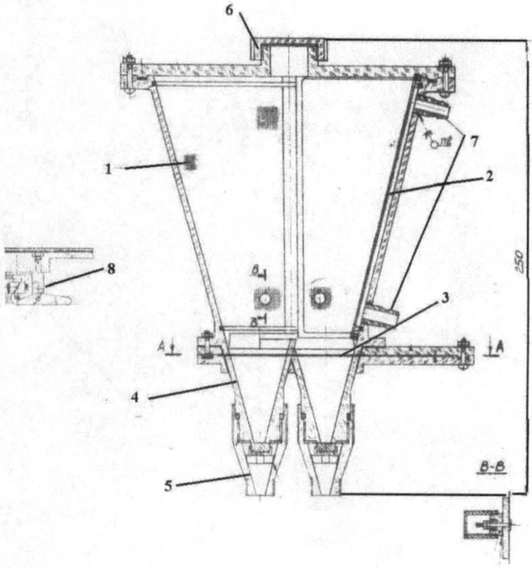

Figure 12: Universal electric separator-concentrator
1, 2 - metal mesh electrodes; 3 - shutter with louver mechanism;
4 - sampling outlet cocks; 5 - receiving collector capsules;
6 - cover; 7 spigots for flushing the separator after completion of work;
8 - shutter release mechanism.

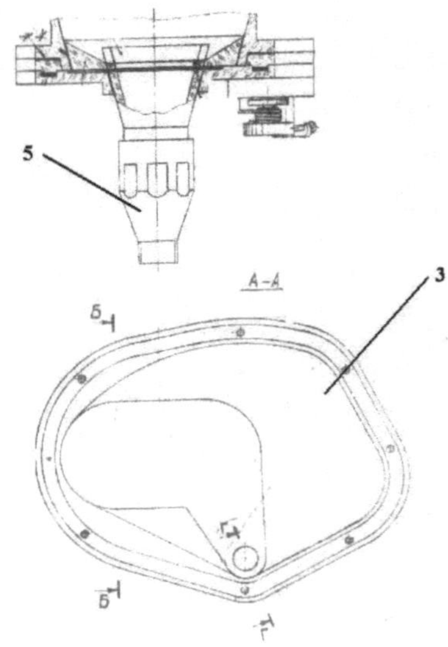

Figure 13: Gate with louver mechanism (3) and receiving manifold capsule (5)

The universal electric separator of conical shape contains two metal mesh electrodes (1, 2), shutter with louver mechanism (3), sampling outlet taps (4), receiving collector capsules (5), lid (6), spigots for washing the universal separator after the end of operation (7) and shutter release mechanisms (8).

The conical chamber of the universal separator is made of organic glass.

Metal mesh electrodes made of non-corrosive material are rigidly fixed on the walls of the universal electric separator with the help of a special frame.

The curtain mechanism and brass gate valve are located downstream of the chamber and are designed to separate the aqueous mixture in the conical chamber from the outlet taps.

The separator outlet taps are cone-shaped and are used to sample clean water and solid micro-particles or microalgae cells.

The tests of the universal electric separator on board the IL-76K flying laboratory [22] were carried out in the following sequence:

1. At the moment of exit to the "weightlessness" mode n=0;

- the dashboard lights and the movie camera were turned on;

- digital current voltage of 26-36 V was simultaneously applied to the chamber electrodes (1, 2) for formation of microdispersed bubbles hydrogen and fill the entire water mixture with it; the louver mechanism was opened by the gate (3) for mass transfer separation results from the chamber to the outlet cocks (4) and sampling reclaimed water and solid microparticles or cells microalgae into receiving capsules-collectors (5);

2. At the moment of the end of the "weightlessness" mode n= g= 0:

- the power was turned off;

- the lights and the movie camera were turned off.

The tests carried out under short-term weightlessness conditions have shown the reliability of the universal electric separator and the possibility to realize on it the process of water recovery from solid microparticles or microalgae cells in the life support system of the International Space Station.

The footage (Figure 14) captures the operation of the Universal Separator-Concentrator in short-term microgravity conditions.

Round-the-clock operation of the universal separator is ensured by continuous supply of a small voltage to the electrodes of the universal separator from the solar panels of the International Space Station and continuous supply of the source liquid to the universal electric separator.

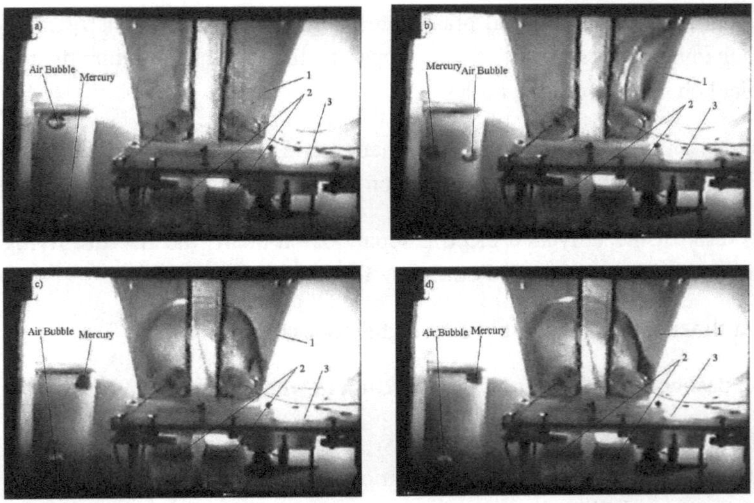

Figure 14: Footage captures the operation of the Universal Separator in microgravity.
a, b, c and d - microgravity in time;
1 - chamber with two metal mesh electrodes;
2 - receiving collector capsules;
3 - shutter with louver mechanism.

CHAPTER 2

24-HOUR RECOVERY OF WATER FROM URINE IN THE LIFE SUPPORT SYSTEM OF THE INTERNATIONAL SPACE STATION IN MICROGRAVITY USING A UNIVERSAL ELECTRIC SEPARATOR

Under weightlessness conditions, under the influence of an electric field on urine (containing 95% water), as on Earth, the formation of hydrogen bubbles, oxygen bubbles and two water molecules occurs as a result of electrolysis. The physical mechanism of formation of negatively charged hydrogen bubbles is described by Shoikhedbrod in [19].

The negatively charged hydrogen bubbles formed can be controlled to a calculated size by electrical potential, making them as small as the process requires.

The use of negative charge and calculated dispersity of hydrogen bubbles allowed to solve the main problem of the separation process: fixation of hydrogen bubbles on solid urine microparticle and formation of strong complexes hydrogen bubbles + urine microparticle.

The physical mechanism for creating this robust complex is as follows.

After formation and detachment from the surface of the cathode, negatively charged hydrogen bubbles moving in urine towards the positively charged anode due to the force of electrostatic attraction in microgravity conditions, meet on their way a urine microparticle with dimensions much larger and induce a positive charge on the surface of the solid particle.

Then, as a result of the attraction of opposite charges, the bubbles adhere to the urine microparticle, forming a firm contact with it (Figure 15).

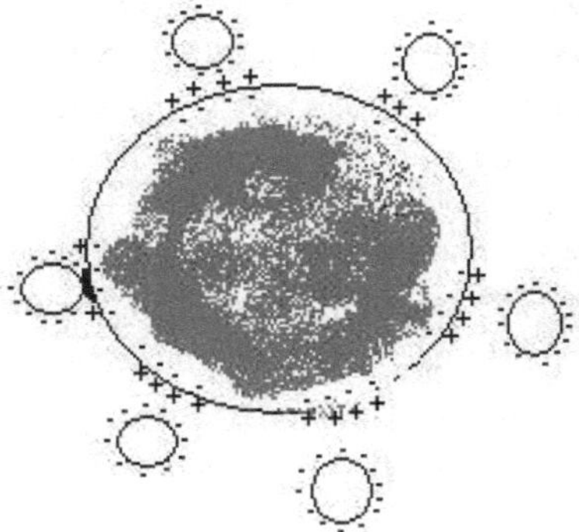

Figure 15: A negatively charged hydrogen bubble induces a positive charge on the surface of a urine microparticle and attaches to it.

The strength of the contact is due to the action of two forces: the surface tension force **$F\sigma = \pi a \sigma Sin(\theta)$** (*a*- radius of the fascination perimeter; σ- surface tension coefficient and *θ*- contact angle and the electrostatic attraction force F_E, which together act to firmly anchor negatively charged hydrogen bubbles on the surface of the urine microparticle

Hydrogen bubble microdispersion allows complete coverage of the entire surface of the urine microparticle if the size of the urine microparticle is much larger than the size of the hydrogen bubble (Figure 16).

As a result, due to the screening of urine microparticles by negatively charged hydrogen bubbles, negatively charged complexes of hydrogen bubbles + urine microparticle are formed.

The formed complexes of hydrogen bubbles + urine microparticle move towards the urine to the positively charged anode due to the force of electrostatic attraction under microgravity conditions.

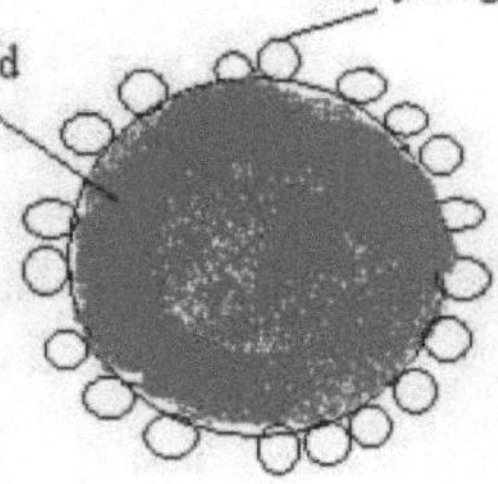

Figure 16: Urine microparticle coated with hydrogen bubbles if the microparticle size is much larger than the hydrogen bubble size

If the size of the urine microparticle is commensurate with the size of the negatively charged hydrogen bubble, the bubble attached to the surface of the particle moves toward the urine toward the positively charged anode for the same reason.

The fact of formation of negatively charged microdispersed hydrogen bubbles in the process of electrolysis, the movement of the formed complexes towards the positively charged anode in urine under microgravity conditions was experimentally proved during flight tests on board the flying laboratory (FL) IL-76K [21].

Figure 17 shows the formation of negatively charged microdispersed hydrogen bubbles and their ascent to the indicated mark as a result of electrostatic attraction of the anode and their retention on the

The system is designed to be used by vibration in short-term microgravity regimes.

In the universal electric separator developed by the author for practical achievement of water recovery from urine in the life support system of the International Space Station in microgravity conditions for mixing negatively microdispersed hydrogen bubbles with urine microparticles and removal of complexes: hydrogen bubbles + urine microparticles the action is used.

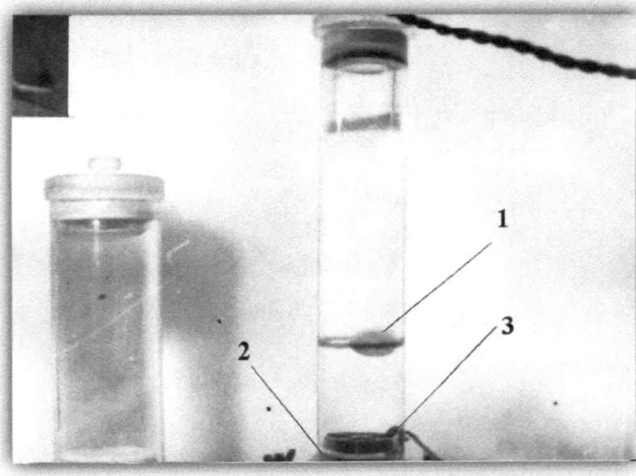

Figure 17: Photograph of the formation of negatively charged hydrogen bubbles during electrolysis and their rise to the indicated mark as a result of electrostatic attraction of a positively charged anode, keeping them at the indicated mark by vibration under short-term weightlessness conditions

1 - Localized negatively charged hydrogen bubbles

2 - Sub-electrode (cathode)

3- Upper electrode (anode)

electrostatic force of the anode and the special conical shape of the electric separator.

In the electroseparator under the influence of the electric field on urine under weightlessness conditions, microdispersed, negatively charged hydrogen bubbles are formed on the cathode. Under the action of electrostatic attraction of the positive anode, hydrogen bubbles fill the entire volume and form strong complexes of microdispersed negatively charged hydrogen bubbles + urine microparticle in the electroseparator.

These complexes will be negatively charged because the negatively charged hydrogen bubbles form a negative screen on the urine microparticle in the complex and will move towards the positively charged anode (separation), accumulating in the area near the anode under weightless conditions.

The lack of Archimedes force creates the problem of delivering complexes (due to the lack of flotation under weightless conditions) to the receiving manifold.

The problem was solved by the design of an electric separator and receiver collector capsules of conical shape.

Under weightlessness conditions, a top-down force equal to $F = \sigma L$ acts on urine contained in a cone-shaped chamber.

From bottom to top, a force equal to $F1 = \sigma L1$ acts on the aqueous mixture. Since $L>L1$, $F>F1$ and hence the resultant force $Fr = F-F1$ will be downward and the resulting urine mixture will flow gently downward as shown in Figure 18.

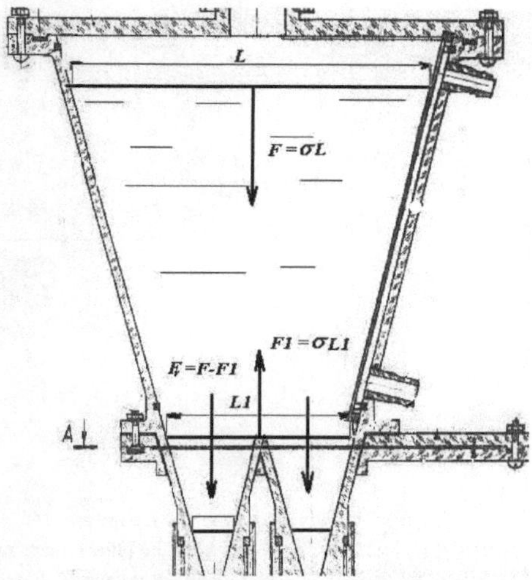

Figure 18: Motion of urine mixture in the universal electricalseparator in the weightlessness conditions

Thus, under weightlessness conditions, one of the receiving collector capsules (anode zone) will be filled with separated complexes of negatively charged microdispersed hydrogen bubbles + urine microparticles, and the other capsule will be filled with reduced purified water.

Figures 19, 20 show a general view of the universal electric separator.

The conical shaped electroseparator contains two metal mesh electrodes (1, 2), a shutter with a louver mechanism (3), sampling outlet taps (4), receiving collector capsules (5), and a cover (6).

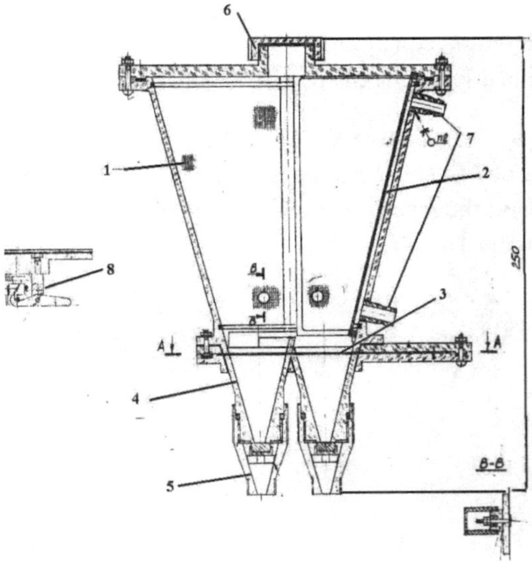

Figure 19: Universal electrical separator
1, 2 - metal grid electrodes; 3- shutter with the blinds mechanism;
4 - selected output crane; 5 - receiving collector capsule; 6 - cover;
7- branch pipes for flushing of separator after the completion of work;
8 - the trigger mechanism of the shutter.

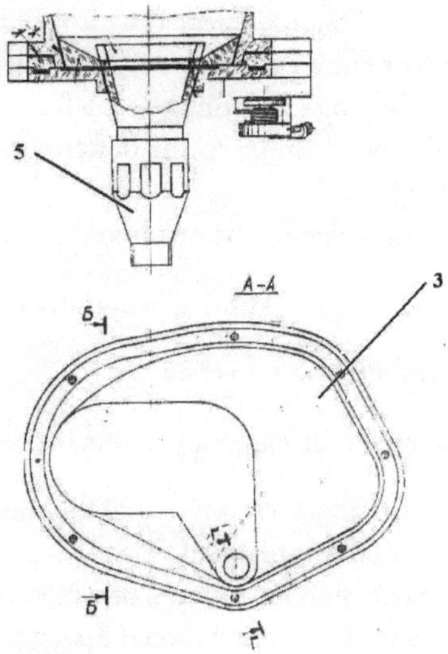

Figure 20: Gate with louver mechanism (3) and receiving manifold capsule (5)

spigots for flushing the separator after the end of operation (7) and gate triggers (8).

The selected separator outlet taps are cone-shaped and serve to withdraw the recovered water and harmful urine inclusions.

The tests of the universal electric separator on board the IL-76K flying laboratory were carried out in the following sequence:

1. At the moment of exit to the "weightlessness" mode n=0;

- the dashboard and movie camera lights were switched on; simultaneously the
camera electrodes (1, 2) were supplied with digital current voltage of 26-36 V
for the formation of microdispersed hydrogen bubbles and the filling of the entire urine mixture and its mixing; the curtain mechanism
was opened by the flap (3) for mass transfer of results
separation from the chamber to the outlet valves (4) and withdrawal of the recovered recovered
water and harmful urine inclusions into the collection capsules (5);

2. At the moment of the end of the "weightlessness" mode n= g= 0:

- the effective voltage was switched off;

- the lights and the movie camera were turned off.

The conducted tests in short-term weightlessness conditions have shown the reliability of the universal electroseparator operation and the possibility to realize on it the process of water extraction from urine in the life support system of the International Space Station.

Thus, the developed universal electroseparator, being simple in application and design, economical and productive, operable in space flight conditions, allows practically realizing water extraction from urine in the life support system of the International Space Station.

Round-the-clock operation of the universal separator is ensured by continuous supply of a small voltage to the electrodes of the universal separator from the solar panels of the International Space Station and continuous supply of urine to the universal electric separator.

CHAPTER 3

ROUND-THE-CLOCK PURIFICATION OF FUEL FLUID IN THE SUPPLY SYSTEM AND TECHNICAL WATER IN THE LIFE SUPPORT SYSTEM OF THE INTERNATIONAL SPACE STATION FROM CONTAMINANTS AND MICROBALANCES UNDER MICROGRAVITY CONDITIONS USING A UNIVERSAL ELECTRIC SEPARATOR

Selection of the obtained pure liquid fuel, purified special liquids of the life support system of the International Space Station in microgravity conditions is carried out in different ampoules due to the special conical shape of the universal electric separator.

Figures 21, 22 show a general view of the universal electric separator.

The universal electric separator of conical shape contains two metal mesh electrodes (1, 2), a shutter with a louver mechanism (3), sampling outlet taps (4), receiving collector capsules (5), a cover (6), spigots for flushing the separator after operation (7) and shutter release mechanisms (8).

The conical chamber of the universal electric separator is made of organic glass.

Metal mesh electrodes made of non-corrosive material are rigidly fixed on the walls of the electric separator by means of a special frame.

The louver mechanism and brass shutter are located downstream of the chamber and are designed to separate the aqueous mixture in the conical chamber from the outlet capsules.

The separator outlet capsules are cone-shaped and are used to extract liquid fuel or water and solid inclusions.

The universal electric separator works as follows.

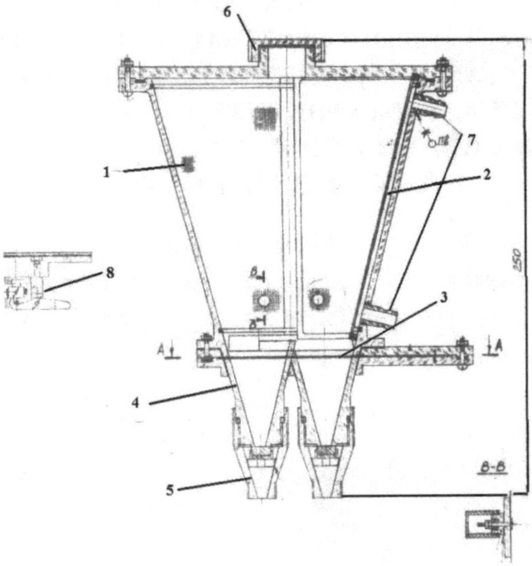

Figure 21: Universal electric separator-concentrator
1, 2 - metal mesh electrodes; 3 - shutter with louver mechanism;
4 - sampling outlet cocks; 5 - receiving collector capsules;
6 - cover; 7 spigots for washing the separator after the end of the

operation

operation; 8 - shutter release mechanism.

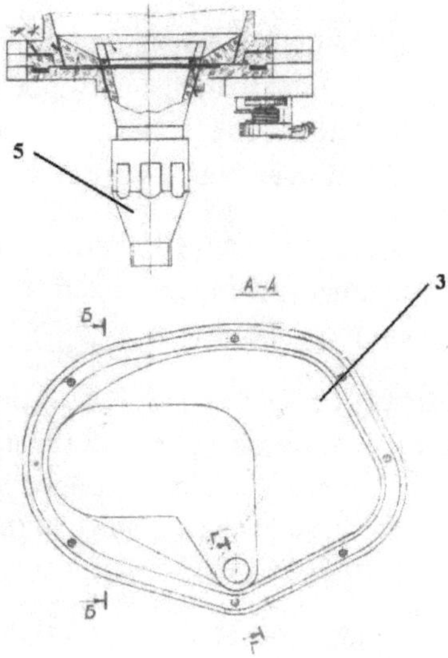

Figure 22: Gate with louver mechanism (3) and receiving manifold capsule (5)

The universal electric separator is pre-filled with the appropriate medium: liquid fuel or water with micro-particles of contaminants or microbial cells.

Due to the electrostatic force between the grid cathode and anode, microdispersed negatively charged hydrogen bubbles formed on the grid cathode move towards the positive grid anode.

Microdispersed negatively charged hydrogen bubbles, meeting a solid microparticle on their way, adhere to it, forming its negative shielding, as a result, the formed complexes microdispersed negatively charged hydrogen bubbles + microparticle move towards the grid anode due to its positive charge.

At the moment of displacement of the largest number of complexes on the grid anode (separation), a gate is opened for withdrawal and extraction of solid microparticles into the capsule (5) on the anode side and the medium purified from solid microparticles into the capsule (5) on the cathode side.

Under microgravity conditions under the action of an electric field, hydrogen and oxygen bubbles and two water molecules are formed on water, just as on Earth. The physical mechanism of formation of negatively charged hydrogen bubbles is described in [19].

The size of the negatively charged hydrogen bubbles produced can be controlled computationally using the electrical potential, making them as small as the process requires.

The use of negatively charged and calculated dispersed hydrogen bubbles allows solving the main problem of the separation process: fixation of hydrogen bubbles on the surface of solid microparticles in the liquid (liquid fuel, water + solid microparticles (contaminant or microbial microparticles),

In terrestrial conditions, the role of mixing of hydrogen bubbles with microparticles, resulting in the formation of complexes: hydrogen bubbles + solid microparticle, as well as the surfacing of the formed complexes is played by Archimedes' force.

In the developed universal electric separator the separation of solid microparticles from the environment in microgravity conditions, the formation of complexes: hydrogen bubbles + solid microparticles is realized by using the action of electrostatic force of attraction of the anode, which makes negative hydrogen bubbles formed at the cathode move towards the anode.

On their way hydrogen bubbles meet with solid microparticles and form complexes hydrogen bubbles + solid microparticles (Figures 23 and 24), which due to the formed negative screen and the action of electrostatic

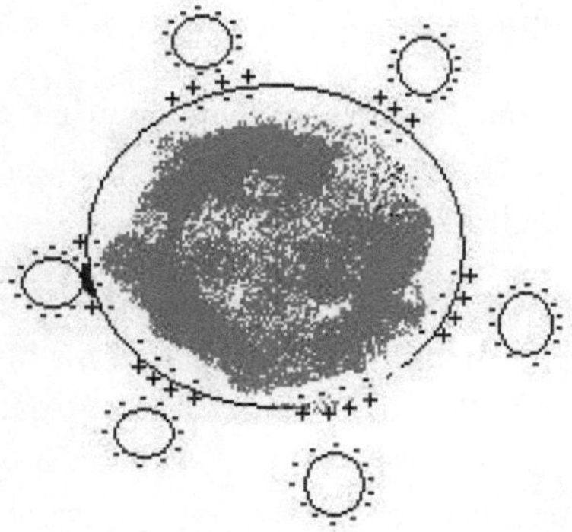

Figure 23: A negatively charged hydrogen bubble inducing a positive charge on the surface of a microparticle in the environment and anchoring on it, resulting in the formation of a negative

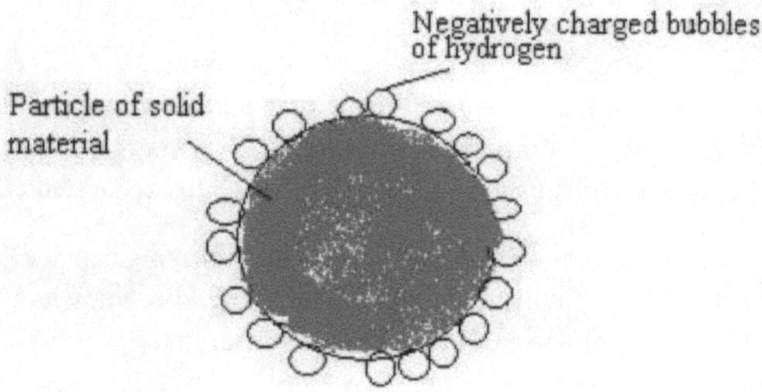

shield.

Figure 24: Ambient microparticle coated with hydrogen bubbles if the size of the microparticle is much larger than the size of the hydrogen bubble, resulting in a negative screen.

of the gravitational attraction shifts toward the anode.

Figure 23 shows a negatively charged hydrogen bubble inducing a positive charge on the surface of a microparticle in the medium and anchoring on it, resulting in a negative screen.

Figure 24 shows a microparticle of a medium covered with hydrogen bubbles if the size of the microparticle is much larger than the size of the hydrogen bubble, resulting in a negative screen.

The separation of complexes of hydrogen bubbles + solid microparticles displaced towards the anode from the purified medium is carried out by moving in microgravity the purified medium and the complexes separated from it into different separated capsules using the conical shape of the separator and the separated capsules.

Under weightless conditions, a top-down force equal to $F = \sigma L$ (σ- surface tension coefficient) is exerted on the purified medium and the complexes isolated from it contained in the conical shape, as shown in Figure 25.

A force equal to $F1 = \sigma L1$ acts from bottom to top.

Since $L > L1$, $F > F1$ and hence the resultant force $Fr = F - F1$ will be directed downward, and the purified medium and the complexes separated from it will flow smoothly downward into the various separated capsules by utilizing the conical shape of the purifier and the separated capsules.

Thus, under weightless conditions, one of the receiving capsule-collectors (on the anode side) will be filled with separated particles and the other capsule (on the cathode side) with purified medium.

In case of purification of liquid propellant of rocket engines and special liquids of the ISS life support system, purified liquid propellant and special liquids act as a purifying medium, and microparticles of contaminants and microbial cells act as solid microparticles.

Tests of the universal electric separator were carried out on board the IL-76K flying laboratory [21].

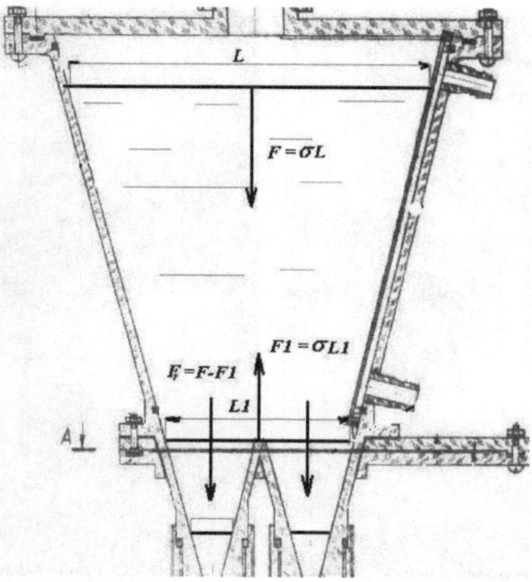

Figure 25: Media movement in a separator-concentrator under microgravity conditions

The conducted tests have shown the reliability of the universal electric separator and the possibility of its practical use for purification of liquid fuel and special liquids from contaminants or microbes in microgravity conditions simply and cost-effectively.

The footage (Figure 26) shows the operation of the universal separator in short-term microgravity

The developed universal electric separator, which uses in its design the effect of direct electric current on liquid fuel, technical water and a special form of construction, makes it possible to clean the fuel liquid from contaminants or microbes in the liquid fuel tank of the ISS (International Space Station) power supply system or special fluids of the ISS life support system for a long time in microgravity conditions simply and economically.

Round-the-clock operation of the universal separator is ensured by continuous supply of small voltage to the electrodes of the universal separator from the solar panels of the International Space Station and continuous supply of dirty liquid fuel

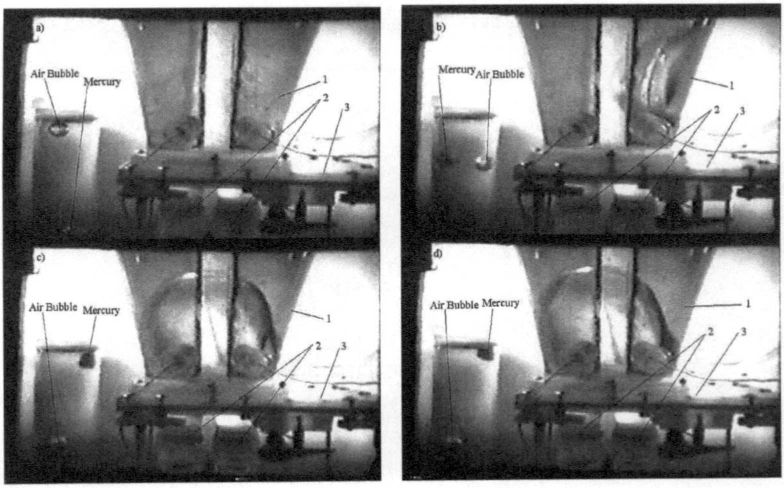

Figure 26: Footage captures the operation of the Universal Separator-Concentrator in microgravity.
a, b, c and d - microgravity in time;
1- chamber with two metal mesh electrodes;
2- Receiving collector capsules;
3- shutter with louver mechanism.

or dirty service liquid into a universal electric separator

CHAPTER 4

ROUND-THE-CLOCK PRODUCTION OF FAST-GROWING MICROALGAE CONCENTRATE FOR CONVERSION OF CARBON DIOXIDE EMISSIONS INTO OXYGEN AND FOOD BIOMASS TO FEED THE CREW OF THE INTERNATIONAL SPACE STATION ON A LONG-DURATION FLIGHT USING A UNIVERSAL ELECTRIC SEPARATOR

Water in a universal electric separator dissociates into ions:

$$H_2O\ H \leftrightarrow ^{+} + OH^{-}$$
(4.1)

When a direct current is passed through water, one hydrogen ion H^+ approaches the cathode, robbing it of an electron, and four OH^- ions approach the anode, which give up four electrons.

As a result, the neutral hydrogen atoms ***H*** formed at the cathode become unstable in the free state and pair up to form the two-atom hydrogen molecule H_2.

Two water molecules are formed at the anode, and the resulting neutral oxygen atoms ***O*** in the free state become unstable and pair up to form the two-atom oxygen molecule O_2.

This will cause hydrogen bubbles to break off at the cathode and oxygen bubbles to break off at the anode:

$$H\ e^{+} + = H; \quad H\ H + = {}_{H(2)}\uparrow$$
(4.2)

$$4OH^{-} - 4e = 2H_2O + {}_{O(2)}\uparrow$$
(4.3)

The physical mechanism for the formation of negatively charged hydrogen bubbles at the cathode and positively charged oxygen bubbles is as follows [19].

A hydrogen molecule in the form of a bubble formed at the cathode is immersed in a cloud of positively charged hydrogen ions H^+ , which induces a negative charge on the outer surface of the bubble, while the inner surface of the bubble will be positively charged.

The negative potential of the cathode will detach the negative charges of the outer surface of the bubble, pushing them to the top of the bubble, and attract the positive charges of the inner surface of the bubble, collecting them in the region near the cathode, as shown in Figure 27 [20].

Figure 27: Photograph of a hydrogen bubble with contact angle θ=97° at the cathode and the catholyte solution (living water) formed around it (magnification x 7.5) [20].

In Figure 2, the sign "+" denotes hydrogen cations H^+ , positive charges of the inner surface of the bubble on the cathode; the sign "-" denotes negative charge - charge of the cathode, negative charges of the outer surface of the bubble.

At the moment of bubble detachment, the positively charged near-cathode part of the bubble remains on the cathode, and the detached hydrogen bubble with negatively charged surface moves upward to the free surface water.

Intensive formation of negatively charged microdispersed electrolysis hydrogen bubbles, their repulsion from the cathode and horizontal movement to the positively charged anode under microgravity conditions saturates water with gaseous hydrogen bubbles, forming negatively charged hydrogen water - catholyte solution (living water).

The neutral oxygen molecule formed at the anode in the form of a bubble is immersed in a cloud of negatively charged ***OH*** hydroxyls$^-$, which induces a positive charge on the outer surface of the bubble, while the inner surface of the bubble will be negatively charged.

The positive potential of the anode, through electrostatic repulsion force, will shift the positive charges on the outer surface of the bubble towards the top of the bubble, and attract the negative charges on the inner surface of the bubble, collecting them in the anode region.

At the moment of bubble detachment, the negatively charged part of the bubble near the anode region remains on the anode, while the detached oxygen bubble with a positively charged surface moves upward to the free surface water.

In this case, the formation of positively charged electrolytic oxygen bubbles, their repulsion from the anode and their movement to the negatively charged cathode saturate the water with gaseous oxygen bubbles, forming positively charged oxygen water - anolyte solution (dead water).

Thus, the bubbles formed at the electrode are always charged equally with the electrode, i.e., hydrogen bubbles take a negative charge and oxygen bubbles take a positive charge.

The bubbles formed on the electrode during electrolysis have different sizes.

The reason for this difference in the size of the bubbles formed on the electrode is down to the different electrostatic interaction of the charged bubbles with the electrode.

The following forces act on the bubble formed in the process of electrolysis and fixed on the electrode surface in the liquid in microgravity conditions: the bubble surface tension force $F\sigma$, by means of which the bubble is held on the electrode surface, acts on the perimeter of the bubble fixing and is equal to $F\sigma = \pi a \sigma Sin(\theta)$, where: **a** = diameter of the circle on which the bubble is fixed on the electrode surface; θ = contact angle at the metal-solution-gas interface; σ = surface tension at the solution-gas interface;

The force of electrostatic interaction of charged bubbles with the electrode F_E.

F_{E1} is the electrostatic force of attraction of the cloud of positive hydrogen ions to the negatively charged outer layer of the hydrogen bubble and the cloud of negative hydroxyl anions to the positively charged outer layer of the oxygen bubble, contributing to the detachment of the bubbles from the electrodes.

Depending on the sign of the charge of the electrode and the bubble, the bubble is either attracted by the electrode (electrostatic force of attraction) or detached from it (electrostatic force of detachment) and, accordingly, reaches a higher or lower value before detachment from the electrode. Then the resulting force acting on the bubble at the points of its contact with the electrode surface can be represented as follows:

$$F_\sigma = F_{E1} + F(E) \quad (4.4)$$

The contributions of F_E and F_{E1} are significant when the bubble detaches from the electrode under microgravity conditions.

Thus, the higher the electrode potential, the smaller the bubble size, because the higher the electrode potential, the greater the electrostatic detachment force acting between the electrode and the bubble formed on the electrode surface of the same charge, detachment of the bubble from the electrode surface, and the faster the bubble detaches.

Thus, the size of the negatively charged hydrogen bubbles can be controlled by increasing the electric potential, making the bubbles as small as necessary for the process.

Studies have shown that the use of negatively charged hydrogen bubbles with calculated dispersity solves the main problem of any technological process associated with the separation of nutrient medium or concentration of cells of microscopic organisms from the nutrient medium: the creation of strong complexes hydrogen bubbles + microalgae cell.

The developed universal electric separator-concentrator concentrates living cells of Chlorella and Dunaliella from their nutrient medium under microgravity conditions, the formation of complexes: negatively charged hydrogen bubbles + Chlorella or Dunaliellacell is carried out using the force of electrostatic attraction of the anode, which makes the negative hydrogen bubbles formed at the cathode move towards the anode.

On their way, the negatively charged hydrogen bubbles meet the Chlorella or Dunaliella cell, which have a positive charge on the outer layer of the phospholipid bilayer of the microalgae cell membrane[22] and form hydrogen bubble + Chlorella or Dunaliellacell complexes, as shown in Figure 28a, 28b due to the action of attraction force between opposite charges and surface tension forces that allow the hydrogen bubble to anchor on the surface of the cell membrane.

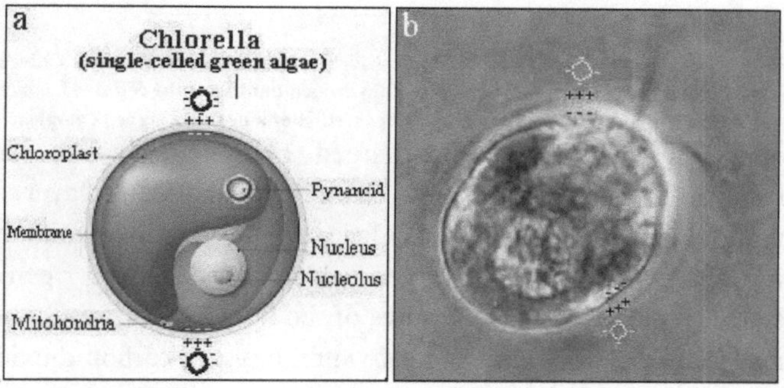

Figure 28: Complex formation: negatively charged hydrogen bubble + Chlorella cell (a) and Dunaliella cell (b) under microgravity conditions

Hydrogen bubble microdispersion allows complete coverage of the entire surface of a Chlorella or Dunaliella cell if the cell size is much larger than the size of the hydrogen bubble.

As a result, the formed complex acquires a negative screen and, due to the resulting action of electrostatic attraction of the anode, shifts towards the anode.

Figure 29 shows the complete coverage of the entire positively charged lipid surface of a Chlorella or Dunaliella cell membrane by negatively charged hydrogen bubbles if the cell size is much larger than the hydrogen bubble size, and the formation of a negative screen complex.

It should be noted that the intensive formation of negatively charged microdispersed electrolytic hydrogen bubbles, their repulsion from the cathode and movement to the positively charged anode in the nutrient medium, saturates this medium with hydrogen gas bubbles, forming negatively charged hydrogen water - catholyte solution (living water).

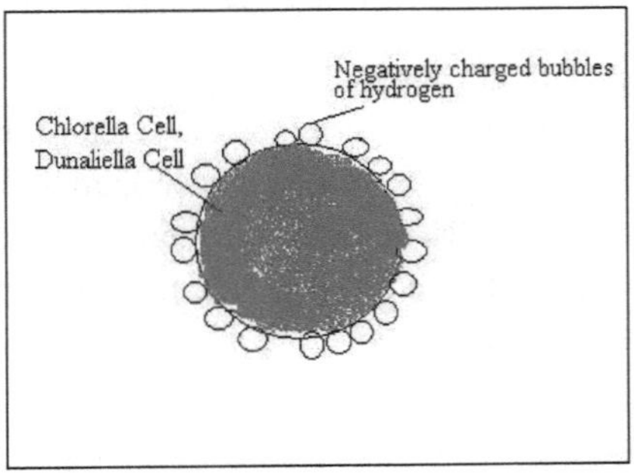

Figure 29: Complete coverage of the entire positively charged lipid surface of a Chlorella or Dunaliella cell membrane by negatively charged hydrogen bubbles if the cell size is much larger than the hydrogen bubble size and the formation of a negative screen complex.

As found in [22], negatively charged hydrogen nutrient medium electrostatically enhances positive charges on the outer layer of the phospholipid bilayer of the microalgae cell membrane while having negative charges on its inner layer, which promotes the opening of cholesterol contained within the pores of the lipid bilayer, resulting in an enhanced increase in the transport of water, oxygen, carbon dioxide and lipophilic substances into the biological cell.

The accelerated release of ***Na***+ 1 cations from the biological cell through the transport protein is accomplished by the electrostatic attraction of the negatively charged catholyte to the released positively charged ***Na***+ 1 cations.

Extracellular ***K+1*** cations bind to the protein, causing the release of a phosphate group, the loss of which restores the original conformation of the protein, and releases **K+1** cations into the cell.

Therefore, living water, being a catholyte, has enhanced biostimulatory properties of microalgae cells.

It has a beneficial effect on Chlorella or Dunaliella cell, activating metabolic processes, increasing its defenses.

Thus, the developed universal electric separator-concentrator utilizes the effect of a constant electric current on the nutrient medium with microalgae, which makes it possible to simultaneously generate microdispersed negative hydrogen bubbles and negatively charged catholyte near the cathode, which, as a result of electrostatic force of attraction between the electrodes, form strong complexes: live microalgae cell + microdispersed hydrogen bubbles concentrated throughout the cathode zone, separated from the microalgae by the cathode.

The conical shape of the vessel containing the complexes: live microalgae cells + microdispersed hydrogen bubbles concentrated along the entire cathodic zone, and the nutrient medium localized along the entire anodic zone, makes it possible to move the separated components into different ampoules under microgravity conditions due to the action of the resulting surface tension force directed downward to the small base of the cone and equal to the difference between the surface tension force applied to the surface of the common mixture of the large base of the cone, and the c

Figures 30, 31 show a general view of a concentrator operating on the previously described principles in microgravity.
The conical-shaped concentrator contains two metal mesh electrodes (1, 2), a shutter with a louver mechanism (3), selected outputs

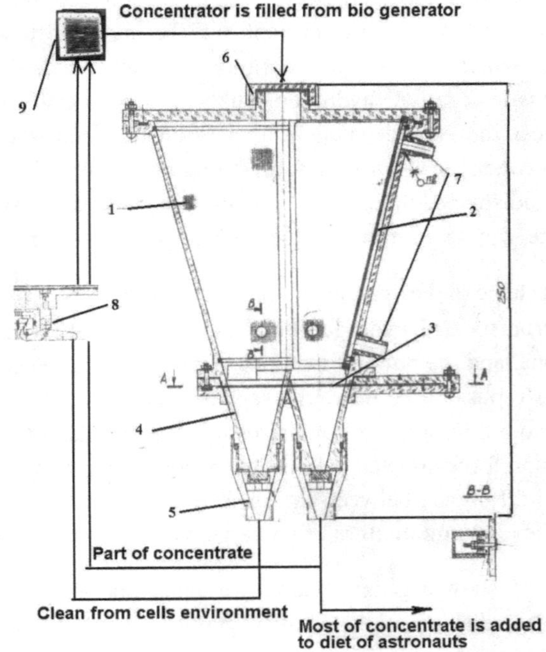

Figure 30: Universal electrical separator-concentrator
1 , 2 - metal mesh electrodes; 3 - Gate with louver mechanism;
4 - Sampling outlet taps; 5 - Receiving manifold capsules;
6 - Cover; 7 - Pipes for flushing the separator after the end of operation operation; 8 - Shutter release mechanism; 9 - Biogenerator.

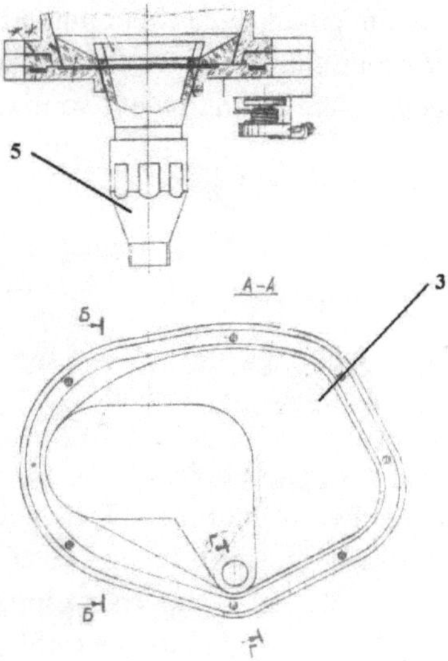

Figure 31: Shutter with louver mechanism (3) and collector receiver capsule (5) bubbles occur.

cocks (4), receiving collector capsules (5), cover (6), spigots for rinsing the separator after the end of operation (7) and gate triggers (8). Conical chamber of the concentrator is made of organic glass.

Metal mesh electrodes made of non-corrosive material are rigidly fixed on the walls of the electroseparator by means of a special frame.

A brass curtain mechanism and shutter, located below the chamber and designed to separate the aqueous mixture in the conical chamber from the outlet capsules.

The individual output capsules of the concentrator are conical in shape and serve to select microalgae cell-free nutrient medium and high concentrate of live microalgae cells for further use on the ISS: all of the microalgae cell-free nutrient medium and a small portion of the high concentrate of live microalgae cells is returned to the photobiogenerator to maintain microalgae cultivation for extended periods of time in a closed loop; most of the high concentration of live microalgae cells

The concentrator operates as follows.

The concentrator is filled with Chlorella or Dunaliella cells and their nutrient medium from the photobiogenerator (9) in a closed loop.

At this point, the separator gate (3) closes, separating the media from the receiving collector capsules (5).

When an electrical voltage is applied to the electrodes of the metal grid (1, 2) containing a mixture of Chlorella or Dunaliella cells or Dunaliella cells and their nutrient medium, in the range of 26-36 V, electrolysis of the Chlorella or Dunaliella cells and their nutrient medium occurs and microdispersed hydrogen bubbles are formed.

Due to the electrostatic force between the grid cathode and anode, microdispersed negatively charged hydrogen bubbles formed on the grid cathode move towards the positive grid anode.

Microdispersed negatively charged hydrogen bubbles, meeting a solid microparticle on their way, stick to it, forming its negative screen, as a result of which the formed complexes microdispersed negatively charged bubbles hydrogen + microalgae cell move towards the reticulated anode due to its positive charge.

At the moment of displacement of the highest number of complexes on the reticulated anode (detachment), the photocell signals the control unit and the control unit opens the gate to withdraw and select a high concentration of Chlorella or Dunaliella cells into the capsule (5) on the anode side and nutrient-free Chlorella or Dunaliella cells into the capsule (5) on the cathode side.

Under weightlessness conditions, a top-down force equal to $\boldsymbol{F} = \boldsymbol{\sigma L}$ (σ-surface tension coefficient) acts on the free surface of the total mixture contained in a conical-shaped concentrator.

A force equal to $\boldsymbol{F1} = \boldsymbol{\sigma L1}$ acts from bottom to top (Figure 32).

Since $\boldsymbol{L > L1}$, $\boldsymbol{F > F1}$ and hence the resulting force $\boldsymbol{Fr = F - F1}$ will be directed downward and the selected high concentration of Chlorella or Dunaliella cells and the selected Chlorella or Dunaliella cell free nutrient medium will flow smoothly into different selected capsules using the conical shape of the concentrator and the selected capsules.

Thus, under weightless conditions, one of the receiver capsules (anode side) will be filled with a high concentration of Chlorella or Dunaliella cells, and the other receiver capsule will be filled with Chlorella or Dunaliella cell-free nutrient medium.

All of the nutrient medium purified from microalgae cells and a small portion of the high concentration of live microalgae cells is returned to the photobiogenerator to maintain microalgae cultivation for extended periods of time in a closed loop; The majority of the high concentration of live microalgae cells is added to the astronauts' diet during extended flight, as shown in Figure 30.

At the moment when the total liquid mixture flows out of the chamber, the photocell signals the control unit and the control unit closes the gate (5), allowing the chamber to fill with microalgae cells and their nutrient medium from the photobiogenerator, and the process thus starts a new cycle.

Tests of the concentrator to confirm the performance of the developed method of obtaining concentrate of rapidly multiplying microalgae in microgravity were conducted on board the IL-76K flying laboratory [21].

The movie footage in Figure 32 captures the operation of the concentrator in microgravity.

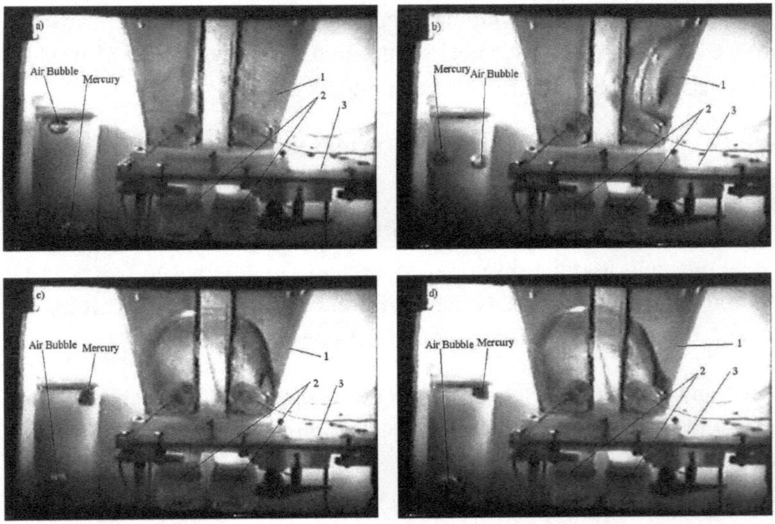

Figure 32 a, b, c, d: Frames captured the operation of the concentrator in microgravity.
1 - Chamber with two metal mesh electrodes;
2 - Receiving collector capsules;
3 - Shutter with louver mechanism.

The conducted tests showed the reliability of the concentrator operation in microgravity conditions and the possibility of its practical use to replenish the photobiogenerator with the concentrate of rapidly multiplying microalgae cells and its nutrient medium in order to grow enough biomass for a long flight, which allows to maintain a normal atmosphere on the ISS by utilizing carbon dioxide and generating oxygen in it.

The concentrate of rapidly multiplying microalgae cells obtained by the concentrator in sufficient quantity and used as a high-calorie supplement in the astronauts' diet allows maintaining the astronauts' performance during long space expeditions.

Thus, the developed universal electric separator-concentrator utilizes the effect of direct electric current on the nutrient medium with microalgae, which allows the simultaneous generation of microdispersed negatively charged hydrogen bubbles and negatively charged catholyte near the cathode, which, as a result of electrostatic force of attraction between the electrodes, form strong complexes: live microalgae cells + microdispersed hydrogen bubbles concentrating throughout the cathode zone, separated from the microalgae by the microalgae.

The conical shape of the vessel containing complexes: live microalgae cells + microdispersed hydrogen bubbles concentrated throughout the cathodic zone, and nutrient medium localized throughout the anodic zone, makes it possible to move the separated components into different ampoules under microgravity conditions.

The separated nutrient medium and a portion of the live microalgae cell concentrate are returned to the cultivator via a closed loop to further grow microalgae, remove carbon dioxide, and replenish oxygen in the ISS atmosphere during the long flight.
Most of the remaining concentrate is used to feed astronauts for their performance during a long flight.
The conducted tests have shown the reliability of the concentrator in microgravity conditions and the possibility of its practical use to replenish the photobiogenerator with the concentrate of rapidly multiplying microalgae cells and its nutrient medium in order to grow enough biomass for a long flight, which allows to maintain a normal atmosphere on the ISS by utilizing carbon dioxide and generating oxygen in it.

The concentrate of rapidly multiplying microalgae cells obtained by the concentrator in sufficient quantity, which is used as a high-calorie supplement in the astronauts' diet, makes it possible to maintain the astronauts' working capacity during long space expeditions.

Round-the-clock operation of the universal separator=concentrator is provided by continuous supply of small voltage to the electrodes of the universal separator-concentrator from solar batteries of the International Space Station and continuous supply of nutrient medium to microalgae cells in the universal electric separator-concentrator.

CHAPTER 5

ROUND-THE-CLOCK PRODUCTION OF HYDROGEN AND OXYGEN GAS FOR CONTINUOUS CHARGING OF FUEL ELEMENTS TO SUPPLY THE POWER SYSTEM OF THE INTERNATIONAL SPACE STATION IN MICROGRAVITY USING A UNIVERSAL ELECTRIC SEPARATOR

Today on the International Space Station (ISS), the Russian module is equipped with a modernized system "Elektron-VM" for the production of oxygen and hydrogen based on electrolysis of water containing liquid alkali, with increased capacity (80...160 l/h), which allows to provide oxygen for up to six people [23].

The Electron-VM Oxygen Generation System (OGS) utilizes water collected from various sources within the ISS (e.g., urine, technical wastewater, and condensate) to produce hydrogen ($H_{(2)}$) and oxygen (O_2) by water electrolysis.

Figure 33 shows a schematic of the electrolysis process used in the Electron-VM OGS system in one of several electrolyzers.

Potassium hydroxide (***KOH***) is used as the electrolyte to create a solution containing 30% ***KOH*** with water.

When current is passed through such a solution, oxygen and hydrogen are produced.

Oxygen and hydrogen are realized through static separators, with oxygen entering the ISS atmosphere and hydrogen entering the Sabatier reactor, where H_2 and CO_2 are combined to create water and methane as follows:

CO_2 (crew respiration) + $2H_{(2)}$+ $2H_{(2)}$(electrolyzed hydrogen is added twice) $\rightarrow 2H_2O + CH4$ (5.1)

Methane CH_4 is converted to carbon and hydrogen as follows:

$$CH_4 + \text{heat} \rightarrow C + 2H_2 \qquad (5.2)$$

The water is returned to the OGS, completing the ISS life support regeneration cycle, while the methane is released into space. The ISS OGS consumes about 23 liters of water per day at a constant current of 50 amperes of each electrolytic cell in the OGS system.

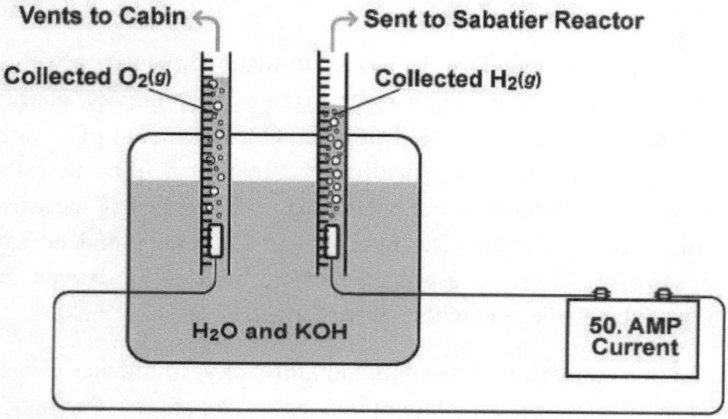

Figure 33: Schematic representation of the electrolysis process used in the Electron-VM OGS system in one of several electrolytic cells.

The oxygen produced fully meets all the requirements for oxygen used for human respiration.

The OGS water electrolysis system, delivered to the ISS in 2008 [24-26], operates in the U.S. ISS module to produce oxygen.

The system is based on solid polymer electrolytes, which is an advantage, but requires additional water treatment.

The electrodes in the electrolyzer of the OGS system "Electron-VM" or in the OGS system operating in the American module of the ISS have a minimum area so that, according to the idea of the system developers, "run off" from the electrodes bubbles of hydrogen and oxygen in the absence of Archimedes' force.

Therefore, for outgassing, an alkaline electrolyte (30% by weight ***KOH*** solution) is added to the water to increase the electrical conductivity of the water and consequently increase the current density (the current reaches 50 A), which leads to an increase in outgassing mass.

The main disadvantage of this method of obtaining gaseous oxygen and hydrogen in microgravity is the huge energy consumption, the use of ***KOH*** solution or solid polymer electrolytes, which makes it uneconomical, difficult to use and not harmless.

The previously developed mechanism of formation of negatively charged electrolytic bubbles of hydrogen and oxygen in microgravity, confirmed experimentally, allowed to establish horizontal movement of negatively charged electrolytic bubbles of hydrogen in the liquid from the cathode, where they were formed, to the anode due to the emerging electrostatic force of attraction between the bubbles and the anode, and horizontal movement of positively charged electrolytic bubbles of oxygen in the liquid from the anode, where they were formed

As a result, a concentrated gas-liquid mixture: oxygen bubbles + water is formed in the anode zone along the entire length of the anode; a concentrated gas-liquid mixture: oxygen bubbles + water is formed along the entire length of the cathode.

Due to the conical shape of the developed universal electroseparator and its outlet openings under microgravity conditions, the separation discharge of two gas-liquid mixtures into different ampoules is realized.

These two different ampoules, equipped with static separators, make it possible to produce gaseous oxygen and hydrogen separately by electrolysis and to supply hydrogen and oxygen continuously to fuel cells during space flight on the International Space Station.
The essence of the mechanism of formation of charged electrolysis bubbles of oxygen and hydrogen on the electrodes during electrolysis in microgravity is as follows.

Just as in terrestrial conditions under the influence of an electric field on water in microgravity conditions, water dissociates into H^+ and OH^- ions.

When a direct current is passed through water, the hydrogen ion H^+ approaches the cathode, taking an electron from it, and four OH^- ions approach the anode, giving up four electrons.

As a result, the neutral hydrogen atoms ***H*** formed at the cathode become unstable in the free state and combine into pairs of the two-atom hydrogen molecule $\boldsymbol{H_2}$.

Two water molecules are formed at the anode, and the resulting neutral oxygen atoms O in the free state become unstable and pair up to form the two-atom oxygen molecule $\boldsymbol{O_2}$.

Thus, hydrogen bubbles will bounce off at the cathode and oxygen bubbles will bounce off at the anode.

The formation of negatively charged hydrogen bubbles at the cathode and positively charged oxygen bubbles occurs as follows.

A hydrogen molecule in the form of a bubble formed at the cathode is immersed in a cloud of positively charged hydrogen ions $\boldsymbol{H^+}$, which induces a negative charge on the outer surface of the bubble, while the inner surface of the bubble will be positively charged.

The negative potential of the cathode strips away the negative charges of the outer surface of the bubble, pushing them to the top of the bubble surface, and attracts the positive charges of the inner surface of the bubble, collecting them in the near-cathode region (Figure 34[20]).

At the moment of bubble detachment, the positively charged near-cathode part of the bubble remains on the cathode, and the detached hydrogen bubble with

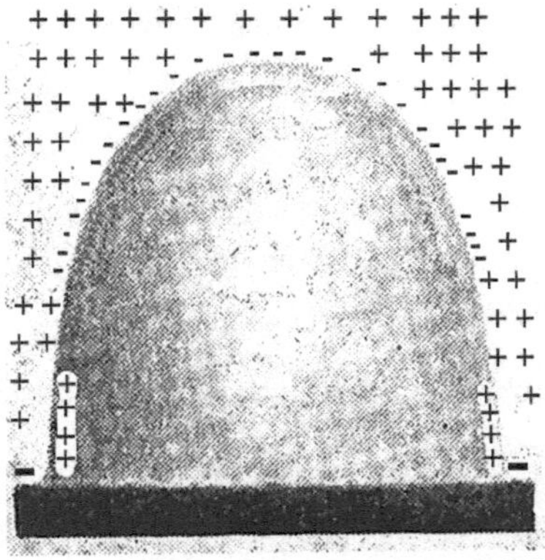

Figure 34: Photograph of a hydrogen bubble with contact angle θ=97° at the cathode (7.5X magnification) [10].

"+" - hydrogen cations *H+*, positive charges of the inner surface of the bubble on the cathode;
"-" - cathode charge, negative external charges of the bubble surface

negatively charged surface breaks away and moves upward toward the free surface of the water.

The neutral oxygen molecule formed at the anode in the form of a bubble is immersed in a cloud of negatively charged OH^-hydroxyls, which induces a positive charge on the outer surface of the bubble, while the inner surface of the bubble will be negatively charged.

The positive potential of the anode detaches positive charges from the outer surface of the bubble, pushing them to the top of the bubble surface, and attracts negative charges from the inner surface of the bubble, collecting them in the anode region.

At the moment of bubble detachment, the negatively charged anodic part of the bubble remains on the anode, while the oxygen bubble with a positively charged surface detaches and moves upward to the free surface of water.

Thus, the bubbles formed on the electrodes are always charged in the same way as the electrode, i.e. hydrogen bubbles acquire a negative charge and oxygen bubbles acquire a positive charge. This conclusion was confirmed during tests conducted on board the IL-76K flying laboratory.

The above theoretical conclusion, confirmed experimentally, about the positive charge of electrolytic oxygen bubbles and negative charge of electrolytic hydrogen bubbles and their detachment from the electrodes, on which they are fixed in the liquid, in microgravity conditions, allows us to assume the principal possibility of horizontal movement of negatively charged electrolytic hydrogen bubbles in the liquid from the cathode, where they were formed, to the anode due to the emerging electrostatic force of attraction between the bubble

At the same time, a concentrated gas-liquid mixture: hydrogen bubbles + water will be formed in the anode zone along the entire length of the anode; a concentrated gas-liquid mixture: oxygen bubbles + water will be formed along the entire length of the cathode.

The conical shape of the developed universal electric separator and its outlet openings made it possible to perform separate discharge of two gas-liquid mixtures into different ampoules under microgravity conditions.

These two different ampoules, equipped with static separators, allow gaseous oxygen and hydrogen to be produced separately by electrolysis and continuously supply hydrogen and oxygen to fuel cells during a space mission on the International Space Station.

The separate discharge of two gas-liquid mixtures into different ampoules under microgravity conditions, due to the conical shape of the developed universal electric separator and its outlet openings, can be explained as follows.

Under microgravity conditions, the resultant force acting on the aqueous medium in the conical chamber is equal to the difference of the surface tension forces acting along the large base of the cone, equal to $F = \sigma L$ (where σ is the surface tension coefficient) and directed from the top to the bottom of the chamber, and the surface tension forces acting along the large bases of the receiving conical capsules, equal to $F1 = \sigma L1$ and directed from the bottom to the top of the chamber, as shown in Figure 35.

Since $L > L1$, $F > F1$ and hence the resulting force $Fr = F - F1$ will be directed downward and the resulting gas-liquid mixtures: electrolyzed hydrogen bubbles + water and electrolyzed oxygen bubbles will flow smoothly downward into the various selected capsules

The design of the developed universal electric separator is as follows.

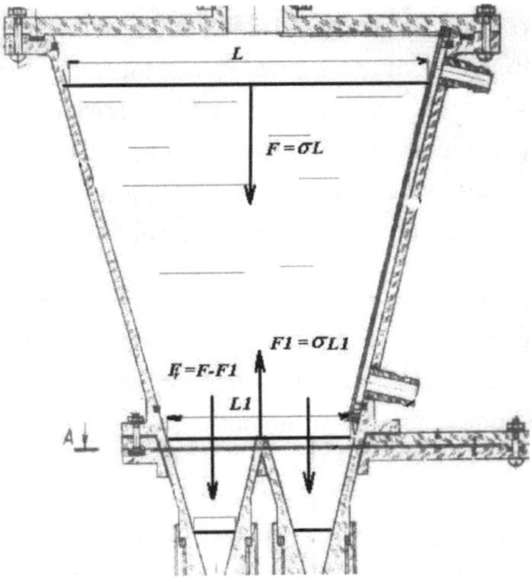

Figure 35: Medium motion in a universal electric separator under microgravity conditions

Figures 36, 37, 38 show the general view of the developed universal electroseparator and two receiving capsule-collectors (1- Figure 37, 38) and two ampoules (2- Figure 37, 38) equipped with static separators to supply the obtained oxygen (Figure 37) and hydrogen (Figure 38) to the fuel cell.

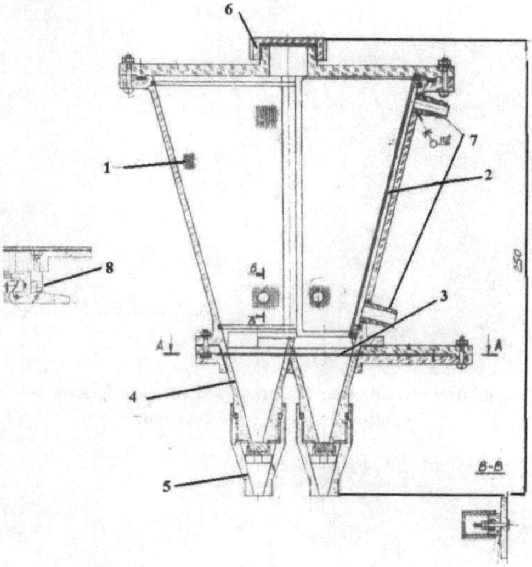

Figure 36: Developed universal electric separator
1, 2 - metal mesh electrodes; 3 - shutter with louver mechanism;
4 - sampling outlet cocks; 5 - receiving manifold capsules;
6 - cover; 7 spigots for washing the separator after the end of the
operation

operation; 8 - shutter release mechanism.

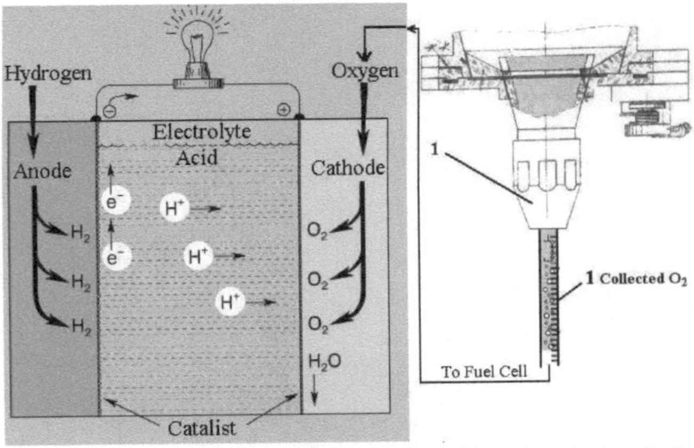

Figure 37: Receiving ampoule-collector (1), equipped with static separators, feeding the resulting oxygen to the fuel cell.

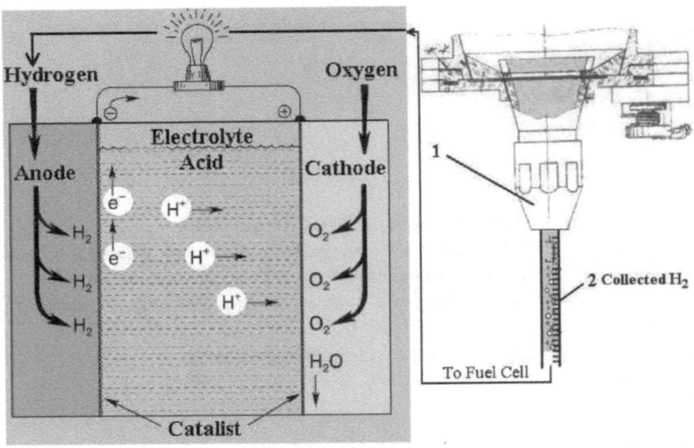

Figure 38: Receiving ampoule-collector (2) equipped with static separators, feeding the produced hydrogen into the fuel cell.

The conical universal electroseparator contains two metal mesh electrodes (1, 2); a shutter with closing spring mechanism (3); two conical outlet valves (4); two conical receiving capsules-collectors (5- Figure 36 or 37 Figure 37, 38) and two ampoules (2- Figure 37, 38) equipped with static separators; cover of the universal electroseparator (6), covering the opening through which water enters the universal electroseparator during continuous operation of the universal electroseparator, outlet outlets for washing the mechanism (3); two conical outlet taps (4); two conical receiving collector capsules (5- Figure 8 or 1- Figure 37, 38) and two ampoules (2- Figure 37, 38) equipped with static separators; universal electroseparator cover (6) closing the opening through which water enters the universal electroseparator during continuous operation of the universal electroseparator, outlet outlets for flushing the universal electroseparator at the end of operation (7) and a spring mechanism closing the shutter (8).

The conical chamber of the universal electroseparator is made of organic glass. Metal mesh electrodes made of non-corrosive material are rigidly fixed on the walls of the universal electroseparator by means of a special frame.

The louvers and brass shutter are located below the chamber and are designed to introduce the resulting separate two separated gas-liquid mixtures (GLMs): hydrogen + water electrolysis bubbles and oxygen + water electrolysis bubbles into two different conical outlet taps and then into two different outlet conical capsules, and finally into two different ampoules fitted with static separators for separate hydrogen and oxygen gases (Figures 37, 38).

The developed universal electric separator works as follows.

The top of the universal electroseparator is filled with reclaimed potable water from atmospheric moisture condensate or urine [28].

At the same time, the gate (3) of the universal electroseparator separating the chamber contents - water from the receiving collector capsules (5) closes.

When an electric voltage is applied to the metal grid of electrodes (1, 2) in the range of 26-36 V, water electrolysis occurs and the formation of microdispersed electrolytic bubbles of hydrogen, negatively charged, and oxygen, positively charged.

Due to the force of electrostatic attraction acting between the cathode grid and positively charged oxygen, the bubbles formed at the anode are displaced horizontally to the near-cathode zone along the entire surface of the cathode grid, forming a gas-liquid mixture: electrolyzed oxygen bubbles + water.

Due to the electrostatic attraction force acting between the anode grid and the negatively charged hydrogen, the bubbles formed at the cathode are displaced horizontally to the anode zone along the surface of the entire anode grid, forming a gas-liquid mixture: electrolyzed hydrogen bubbles + water.

Due to the conical shape of the developed universal electroseparator of its two capsules and two ampoules, the gas-liquid mixtures formed: electrolytic oxygen bubbles and electrolytic hydrogen bubbles + water will flow smoothly into different receiving collector capsules, while the gate (3) of the universal electroseparator separating the contents of the chamber - water with gas-liquid mixtures from the receiving collector capsules (5) will be open.

Thus, under microgravity conditions, one of the receiving collector capsules (on the anode side) will be filled with a gas-liquid mixture: electrolyzed hydrogen bubbles + water, and the other capsule (on the cathode side) will be filled with a gas-liquid mixture: electrolyzed oxygen bubbles + water.

After passing the gas-liquid mixtures: electrolytic hydrogen bubbles + water and electrolytic oxygen bubbles + water through a static separator, the resulting oxygen and hydrogen gases are fed into the spacecraft fuel cell to continuously supply hydrogen and oxygen to the spacecraft fuel cells during a long space mission (Figures 37,38).

In case of continuous operation of the universal electric separator, the lid of the chamber is open all the time and the shutter is automatically closed to fill the chamber with water and complete saturation of the cathode and anode with separate gas-liquid mixtures in the process of electrolysis, at the moment of complete saturation the photocell gives a signal to the control unit, by means of which the shutter is automatically opened.

After the oxygen and hydrogen gases have been produced from all the water contained in the chamber, the photocell signals the control unit and the shutter closes and the process is thus repeated many times.

It should be noted that the gas content of the universal electric separator system determines the efficiency of the universal electric separator.

According to Faraday's law, the mass of gas (***M***) released at the electrode during electrolysis at time t will be:

$$M = jigSt \tag{5.3}$$

where j - electrochemical equivalent of gas (for hydrogen j = 1.0-10-7 kg/(A· s); for oxygen j = 8.29-10-8 kg/(A· s)); i - current density, A/m^2; g = 90%, 95% - current yield (ratio of actual gas mass transfer to theoretical, calculated in percent); S - area of the electrode, m^2, on which hydrogen or oxygen bubbles are released.

From the formula (5.3) it is well seen that when the area of the electrode on which bubbles are released during electrolysis is small, the mass of the released gas (hydrogen and oxygen bubbles) is minimal.

Figure 33 clearly shows that the electrodes in the electrolysis cell of the Electron-VM OGS system have a minimum area.

Electrodes in the electrolysis cell of the "Electron-VM" OGS system have a minimum area so that, according to the idea of the system developers, hydrogen and oxygen bubbles "flow" from the electrodes in the absence of Archimedes' force.

Therefore, for gas release, an alkaline electrolyte (25% by weight ***KOH*** solution) is added to the water to increase the electrical conductivity of the water and hence increase the current density i (current up to 50 A), which leads to an increase in the mass of gas release (***M***).

The developed universal electric separator uses the charge of released electrolytic bubbles of hydrogen and oxygen and the action of repulsive and attractive electrostatic forces of the positive anode and negative cathode during electrolysis in microgravity conditions to obtain intensive mass release of electrolytic gas bubbles of hydrogen and oxygen and therefore does not require the addition of electrolyte in water and at a voltage on the grid electrodes of 23 V the current does not exceed 0.5 A.

In addition, the developed universal electric separator in microgravity conditions due to conical shapes: chamber of the universal electric separator, special outlet valves and capsule-collectors separately directs gas-liquid mixtures: electrolytic bubbles of oxygen + water and electrolytic bubbles of hydrogen + water to different receiving capsule-collectors directly on static separators for separate generation of oxygen and hydrogen gases and their subsequent continuous supply of fuel cells.

Tests of the developed universal electric separator were carried out on board the IL-76K flying laboratory [21].

The tests have shown reliability of the developed universal electroseparator and possibility of its practical use for separate production of gas-liquid mixtures: electrolytic bubbles of oxygen + water and electrolytic bubbles of hydrogen + water and subsequent separate generation of gaseous oxygen and gaseous hydrogen in microgravity conditions simply and economically, which can be used for continuous supply of hydrogen and oxygen to the fuel cells of spacecraft in conditions of prolonged period of time.

The footage (Figure 39) shows the operation of the developed electrolyzer under microgravity conditions.

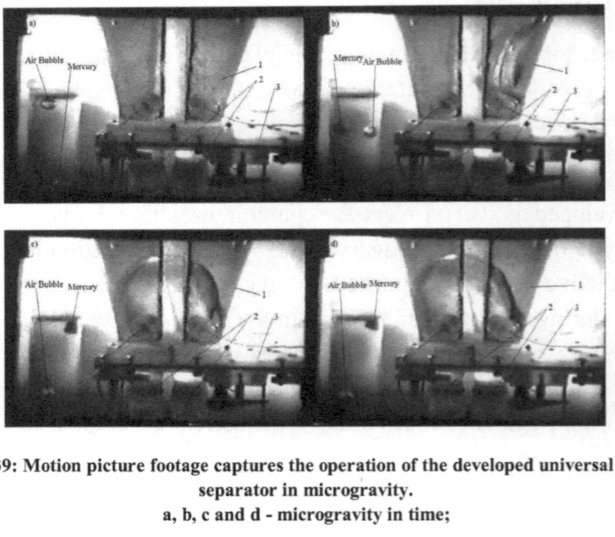

Figure 39: Motion picture footage captures the operation of the developed universal electric separator in microgravity.
a, b, c and d - microgravity in time;

1- a chamber with two metal mesh electrodes;
2- Receiving collector capsules;
3- shutter with louver mechanism.

Thus, the developed method and the universal electroseparator for continuous supply of hydrogen and oxygen to spacecraft fuel cells under conditions of long space flight allow to obtain in the universal electroseparator under microgravity conditions separately concentrated gas-liquid mixtures: negatively charged electrolytic bubbles of hydrogen + water formed in the anode zone along the entire length of the anode; positively charged electrolytic bubbles of oxygen + water formed in the near-cathode zone; positively charged electrolytic bubbles of oxygen + water formed in the anode zone along the entire length of the anode; and positively charged electrolytic bubbles of oxygen + water formed in the near-cathode zone.

The conical shape of the developed universal electroseparator and its output capsules and ampoules allows in microgravity conditions to carry out separate discharge of two gas-liquid mixtures into different ampoules equipped with static separators generating hydrogen and oxygen separately in continuous mode, which can be successfully used for continuous supply of fuel cells during a long space flight.

The developed method and a versatile electric separator allow a safe and economical continuous supply of hydrogen and oxygen to fuel cells.

Round-the-clock operation of the universal separator is ensured by continuous supply of a small voltage to the electrodes of the universal separator-concentrator from the solar panels of the International Space Station and continuous supply of water to the universal electric separator.

CHAPTER 6

ADVANCED LIFE SUPPORT AND POWER SUPPLY SYSTEMS FOR LONG-DURATION MISSIONS ON THE INTERNATIONAL SPACE STATION USING A SYSTEM OF UNIVERSAL ELECTRICAL ISOLATORS

Operation of a universal electroseparator for round-the-clock extraction of water from urine in the life support system of the International Space Station in microgravity conditions; for round-the-clock purification of fuel fluid in the power supply system and technical water in the life support system of the International Space Station from contaminants and microbes in microgravity conditions; for round-the-clock production of a concentrate of fast-growing microalgae for conversion of carbon dioxide into oxygen and food biomass for n

Selection of the obtained pure liquid fuel, drinking water from technically contaminated water and urine, formed as a result of astronauts' life activity, in microgravity conditions is carried out in different ampoules due to the special conical shape of the universal electric separator.

Figures 40, 41 show the separator design in general form.

The conical separator contains two metal mesh electrodes (1, 2), gate with shutter mechanism (3), selective outlet valves (4), receiving collector capsules (5), lid (6), separator flushing spigots after operation (7) and gate release (8).

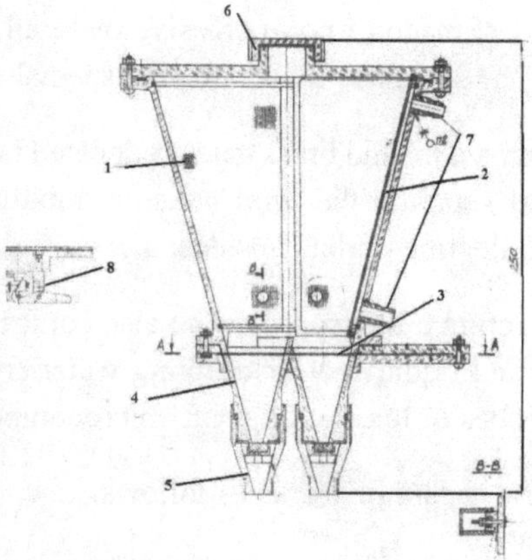

Figure 40: Universal electric separator
1, 2 - metal mesh electrodes; 3 - shutter with louver mechanism; 4 - sampling outlet cocks; 5 - receiving manifold capsules; 6 - cover; 7 spigots for washing the separator after the end of work; 8 - shutter release mechanism

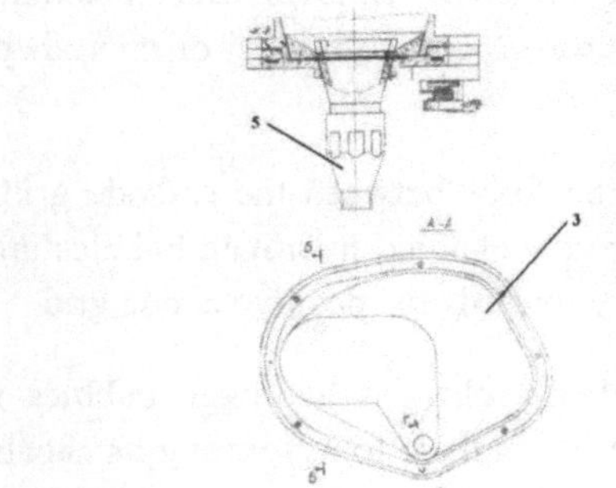

Figure 41: Gate with shutter mechanism (3) and receiving manifold capsule (5)

The conical chamber of the separator is made of organic glass.

Metal mesh electrodes, made of non-corrosive material, are rigidly fixed on the side walls of the separator by means of a special frame.

The gate release mechanism and brass gate are located below the chamber and are designed to separate the total aqueous mixture in the conical chamber from the collecting outlet capsules.

The separator's collecting outlet capsules are also conical in shape and are used to screen purified liquid fuel or drinking water from harmful solid inclusions and microbes or toxic solid urine microcomponents.

The universal electric separator works as follows.

The separator is pre-filled with the appropriate medium: dirty fuel fluid, dirty service water with micro-particles of contaminants or microbial cells or urine.

At the same time, the separator gate (3) is closed to separate the total water mixture from the receiving capsule-collectors (5), by feeding a signal from the photocell located in the lower part of the cathode grid (on the left side of the chamber) to the control unit (transparent color).

When an electric voltage is applied to the metal electrode grids (1, 2) in the range of 26-36 V on a conventional water containing mixture, the process of water electrolysis and formation of microdispersed hydrogen bubbles takes place.

Due to the electrostatic force between the cathode grid and the anode, microdispersed negatively charged hydrogen bubbles are formed on the cathode grid and move towards the positive anode grid.

Microdispersed negatively charged hydrogen bubbles meeting a solid microparticle on their way adhere to it, forming its negative shielding, as a result of which the formed complexes of microdispersed negatively charged hydrogen bubbles + microparticle move towards the anode grid due to its positive charge.

At the moment of accumulation of the largest amount of complexes on the grid anode (separation) and clarification on the grid cathode, the photocell signals the control unit (dark color), which opens the gate for withdrawal and extraction of solid microparticles into the capsule (5) on the anode side and clean fuel liquid purified from solid microparticles or drinking water into the capsule (5) on the cathode side.

In microgravity, under the action of an electric field, hydrogen and oxygen bubbles and two water molecules are formed on water, just like on Earth.

The negatively charged hydrogen bubbles produced can be resized using the electric potential, making them as small as the process requires.

The use of negatively charged calculated dispersion of hydrogen bubbles allows solving the main problem of the separation process: fixation of hydrogen bubbles on the surface of solid microparticles in the liquid (water + solid microparticles (contaminant microparticles or microbial cells), toxic solid microcomponents of urine.

In terrestrial conditions, Archimedes' force ensures the contact of hydrogen bubbles with microparticles, resulting in the formation of complexes: hydrogen bubbles + solid microparticle, as well as flotation of the formed complexes.

In the universal electric separator, the formation of complexes: microdispersed hydrogen bubbles + solid microparticles in microgravity conditions is carried out using the action of the emerging electrostatic force of attraction of the anode, which forces negatively charged hydrogen bubbles formed at the cathode to move to the anode.

Along the way, microdispersed negatively charged hydrogen bubbles meet with solid microparticles and form complexes: microdispersed negatively charged hydrogen bubbles + microparticles of solid material (Figures 42 and 43), which due to the formed negative screen and as a result of the resulting electrostatic attraction force are displaced towards the anode.

Figure 42 shows a negatively charged hydrogen bubble that induces a positive charge on the surface of a microparticle in an aqueous mixture and is anchored to it by the action of electrostatic attraction and surface tension forces, resulting in a negative screen under microgravity conditions.

Figure 43 shows a microparticle in an aqueous mixture coated with microdispersed hydrogen bubbles if the size of the microparticle is much larger than the size of the hydrogen bubble, resulting in a negative screen in microgravity.

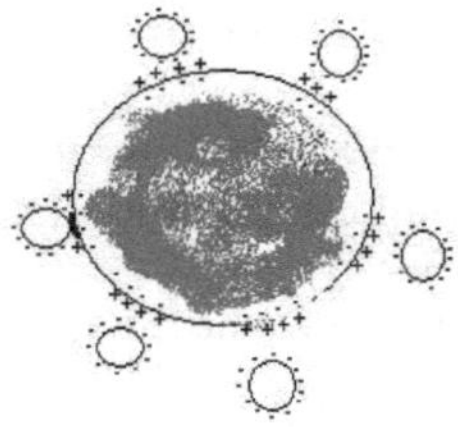

Figure 42: A negatively charged hydrogen bubble inducing a positive charge on the surface of a microparticle in an aqueous mixture and anchoring on it, resulting in the formation of a negative screen under microgravity conditions.

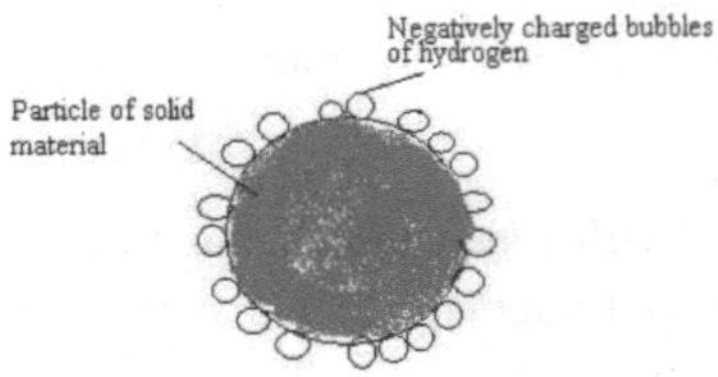

Figure 43: Microparticle of aqueous mixture coated with hydrogen bubbles if the size of microparticles are much larger than the size of a hydrogen bubble, resulting in a negative screen in microgravity.

The selection of the complexes displaced to the anode: hydrogen bubbles + solid microparticles and the purified liquid: clean fuel liquid or drinking water under microgravity conditions is accomplished by moving the purified liquid and the complexes separated from it into different selective capsules under microgravity conditions, utilizing the conical shape of the separator and the separated capsules.

Under weightless conditions, the purified liquid and its separated complexes contained in the conical-shaped separator are subjected to a top-down force equal to $F = \sigma L$ (σ is the surface tension coefficient), as shown in Figure 44.

A force equal to $F1 = \sigma L1$ acts from bottom to top.

Since $L>L1$, $F>F1$ and hence the resultant force $Fr = F-F1$ will be downward and hence the purified liquid and the complexes separated from it will flow smoothly into the various selective capsules using the conical shaped chamber and selective capsules (Figure 44).

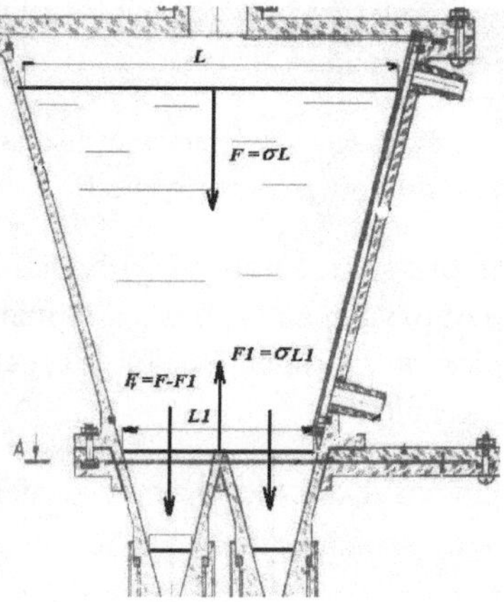

Figure 44: Movement of the separated components of the water mixture

Thus, under weightless conditions, one of the selective capsules (anode side) will be filled with separated particles and the other selective capsule (cathode side) will be filled with purified liquid.

In the case of purification of rocket engine fuel fluid and contaminated technical fluids, purified liquid fuel and drinking water act as purified fluid, and contaminant microparticles and microbial cells act as solid microparticles.

In the recovery of water from urine, the role of purified liquid is played by drinking water recovered from urine, and the role of solid microparticles is played by toxic components of urine.

A universal electric separator-generator has been developed, which makes it possible to economically and simply separately obtain oxygen and hydrogen and use them on the International Space Station (ISS) or on an interplanetary spacecraft for long-term maintenance of the cabin atmosphere with oxygen and fresh water during a long flight, using positively charged oxygen bubbles and negatively charged hydrogen bubbles formed in the process of electrolysis and the effect on them of the electrostatic force arising between the electrodes.

As a result, a concentrated gas-liquid mixture: oxygen bubbles + water is formed in the anode zone along the entire length of the anode; a concentrated gas-liquid mixture: oxygen bubbles + water is formed along the entire length of the cathode.

Under microgravity conditions, a separative discharge of two gas-liquid mixtures into different ampoules is performed.

These two different ampoules equipped with static separators allow the separate production of oxygen and hydrogen in microgravity, opening up new possibilities for the production of oxygen and hydrogen by electrolysis in microgravity.

The design of the universal electric separator-generator is similar to that of the universal electric separator described above (Figure 40).

The only difference is that the designs of the two different output conical capsules (5) in Figure 2 are equipped with static separators for separate hydrogen and oxygen production (Figure 45)

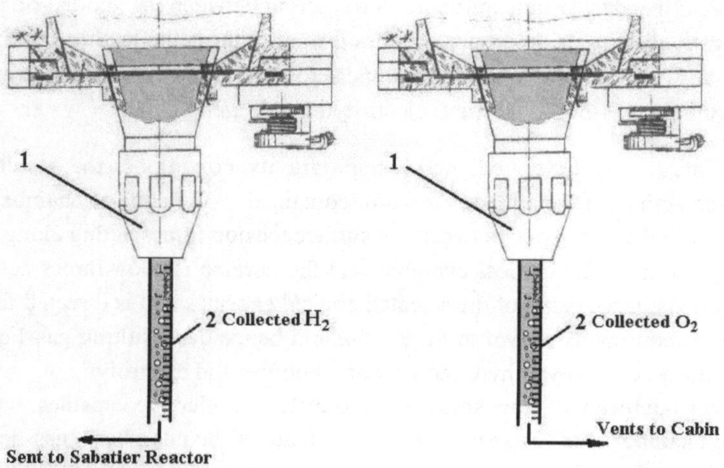

Figure 45: Receiving collection capsules (1) and ampoules (2) equipped with static separators.

The developed universal electric separator-generator works as follows.

The top of the generator is pre-filled with potable water recovered from condensate of atmospheric moisture or urine.

At the same time, the shutter (3) of the universal electric separator-generator separating the contents of the chamber from the capsules (5) is closed by a signal from the photocell located in the lower part of the cathode grid (on the left side of the chamber), directed to the control unit (transparent color).

When an electric voltage in the range of 26-36 V is applied to the metal grids of the electrodes (1, 2), water electrolysis and the formation of microdispersed electrolyzed negatively charged hydrogen bubbles and positively charged oxygen bubbles occur.

Due to the force of electrostatic attraction acting between the cathode grid and positively charged oxygen bubbles formed at the anode, these bubbles horizontally move to the near-cathode zone across the surface of the cathode grid, forming a gas-liquid mixture: electrolytic oxygen bubbles + water.

Due to the electrostatic attraction force acting between the anode grid and negatively charged hydrogen bubbles formed at the cathode, these bubbles move horizontally into the anode zone across the surface of the anode grid, forming a gas-liquid mixture: electrolytic hydrogen bubbles + water.

As previously described, under microgravity conditions, the resultant force acting on the aqueous mixture contained in the conical chamber is equal to the difference between the surface tension forces acting along the large base of the conical chamber and the surface tension forces acting along the large bases of the selected conical capsules and is directed from top to bottom, as shown in Figure 44, and hence the resulting gas-liquid mixtures: electrolytic hydrogen + water bubbles and electrolytic oxygen + water bubbles will flow smoothly into different selective capsules, while the chamber gate (3) separating the contents of the chamber - gas-liquid mixtures from the selective capsules (5) will be opened by the control unit receiving a signal from the photocell (water is turbid due to the gas-liquid mixtures present in it).

Thus, under microgravity conditions, one of the selected capsules (on the anode side) will be filled with a gas-liquid mixture: electrolytic hydrogen bubbles + water, and the other capsule (on the cathode side) will be filled with a gas-liquid mixture: electrolytic oxygen bubbles + water.

After the gas-liquid mixture: electrolytic oxygen bubbles + water passes through the static separator, the resulting oxygen gas enters the astronauts' living quarters and enriches the air breathed by the astronauts with oxygen (Figure 45).

After passing the gas-liquid mixture: electrolytic hydrogen bubbles + water through a static separator, the resulting hydrogen gas enters the Sabatier reactor, where $\boldsymbol{H_2}$ combines with $\boldsymbol{CO_2}$ (astronauts' breath) to form water and methane:

$\boldsymbol{CO_2} +\ \boldsymbol{2H_{(2)}} + \boldsymbol{2H_2}$ *(*electrolyzed hydrogen is added twice*)*$\rightarrow \boldsymbol{2H_2O} + \boldsymbol{CH_4}$ (6.1)

Methane $\boldsymbol{H_4}$ is converted to carbon and hydrogen as follows:

$$\boldsymbol{CH_4} + \textit{heating } C \rightarrow + \boldsymbol{2H_2} \qquad (6.2)$$

The water is then returned to the generator, completing the regenerative life-support cycle on the ISS, while the methane is released into space.

In case of continuous operation of the universal electric separator-generator, the lid of the chamber is open all the time and the shutter is automatically closed to fill the chamber with water and completely saturate the cathode and anode with separate gas-liquid mixtures during the electrolysis process.

At the moment of full saturation, the photocell signals the control unit, which automatically opens the shutter.

After oxygen and hydrogen gases have been produced from all the water contained in the chamber, the photocell signals the control unit (transparent color), the shutter closes, and the process is thus repeated many times.

A universal electric separator-concentrator has been developed that produces a concentrate of rapidly multiplying microalgae under microgravity conditions to utilize carbon dioxide emitted by astronauts and produce oxygen and to add this concentrate as a high-calorie food supplement to the diet of the crew of the International Space Station or interplanetary spacecraft in order to obtain sufficient oxygen, water and algae biomass concentrate for the performance of the crew of the International Space Station or interplanetary spacecraft.

Obtaining the concentrate of rapidly multiplying microalgae under microgravity conditions in a universal electric separator-concentrator is carried out with the help of microdispersed negatively charged hydrogen bubbles formed during electrolysis at the cathode and aqueous solution saturated with them - catholyte, as well as electrostatic force of attraction arising between the electrodes.

The physical mechanism for the formation of negatively charged hydrogen bubbles on the cathode during electrolysis is as follows.

The hydrogen molecule formed at the cathode in the form of a bubble is immersed in a cloud of positively charged hydrogen ions H+, which induce a negative charge on the outer surface of the bubble, while the inner surface of the bubble will be positively charged.

The negative potential of the cathode strips away the negative charges of the outer surface of the bubble, pushing them to the top of the bubble, and attracts the positive charges of the inner surface of the bubble, collecting them in the near-cathode region.

At the moment of bubble detachment, the positively charged near-cathode part of the bubble remains on the cathode, and the detached hydrogen bubble with negatively charged surface moves upward to the free water surface.

Intensive formation of negatively charged microdispersed electrolytic hydrogen bubbles, their repulsion from the cathode and rise to the free surface of water saturates water with hydrogen gas bubbles, forming negatively charged hydrogen water - catholyte solution (living water).

Negatively charged microdispersed microdispersed electrolytic hydrogen bubbles formed on the electrode during electrolysis have microdispersed sizes.

The reason for this microdispersity in the size of the bubbles formed at the electrode is down to the electrostatic interaction of the negatively charged hydrogen bubbles with the cathode.

Under microgravity conditions, the following forces act on the bubble formed in the process of electrolysis and fixed on the electrode surface in the liquid: the bubble surface tension force $F\sigma$, by means of which the bubble is held on the electrode surface, acting along the perimeter of the bubble fixing, and equal to $F\sigma = \pi a \sigma Sin(\theta)$, where: a = diameter of the circle on which the bubble is fixed on the electrode surface; θ = contact angle at the metal-solution-gas interface; σ = surface tension at the solution-gas interface;

The electrostatic force of the electrostatic interaction of negatively charged hydrogen bubbles with the electrode is F_E.

The electrostatic force of attraction of the cloud of positive hydrogen ions to the negatively charged outer layer of the hydrogen bubble and the cloud of negative hydroxyl anions to the positively charged outer layer of the oxygen bubble, which promotes the bubbles to break away from the electrodes F_{E1}.

Thus, since the surface tension force and the electrostatic forces are directed in different directions, there is equality in this case:

$$F_\sigma = F(E) + F(E)(1) \quad (6.3)$$

Thus, under microgravity conditions, the contribution of the forces F_E and F_{E1} is significant to the detachment of the bubble from the electrode.

Thus, the higher the electrode potential, the smaller the bubble size, because the higher the electrode potential, the greater the electrostatic detachment force acting between the electrode and the bubble formed on the electrode surface of the same charge, the faster the detachment of the bubble from the electrode surface.

Thus, the size of the negatively charged hydrogen bubbles can be controlled by increasing the electric potential, making the bubbles as small as necessary for the process.

Negatively charged hydrogen bubbles with calculated dispersity solve the main problem of any technological process associated with the separation of nutrient medium or concentration of cells of microscopic organisms from the nutrient medium: the creation of strong complexes of microdispersed negatively charged hydrogen bubbles + microalgae cell.

In the developed universal electric separator-concentrator the concentration of living Chlorella and Dunaliella cells from their nutrient medium in microgravity conditions, the formation of complexes: negatively charged hydrogen bubbles + Chlorella or Dunaliella cell is carried out with the help of electrostatic attraction force of the anode, which makes the negative hydrogen bubbles formed at the cathode move towards the anode.

On their way, the negatively charged hydrogen bubbles meet with a Chlorella or Dunaliella cell that has a positive charge on the outer lipid side of its membrane and form complexes of microdispersed negatively charged hydrogen bubble + Chlorella or Dunaliella cell, as shown in Figures 46a, 46b, due to the attractive force between the opposite charges and surface tension forces, which in this case are unidirectional, allowing the hydrogen bubble to attach to the surface of the cell membrane.

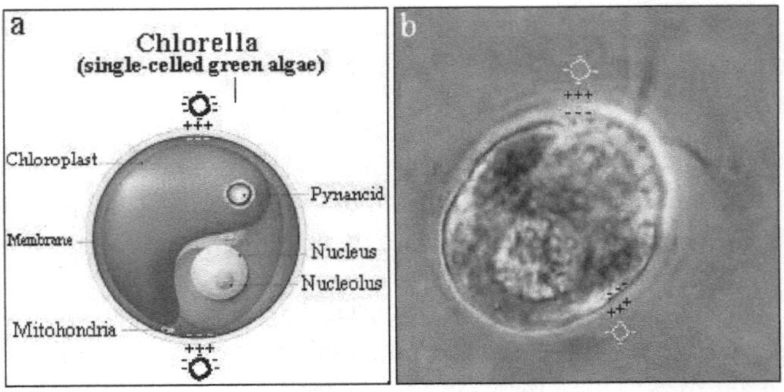

Figure 46 a, b: Complex formation: negatively charged hydrogen bubble + Chlorella cell (a) and Dunaliellacell (b) under microgravity conditions

Hydrogen bubble microdispersion allows complete coverage of the entire surface of a Chlorella or Dunaliella cell if the cell size is much larger than the size of the hydrogen bubble.

As a result, the formed complex acquires a negative screen and, due to the resulting action of electrostatic attraction of the anode, shifts towards the anode.

Figure 47 shows the complete coverage of the entire positively charged lipid surface of a Chlorella or Dunaliella cell membrane by negatively charged hydrogen bubbles if the cell size is much larger than the hydrogen bubble size, and the formation of a negative screen complex.

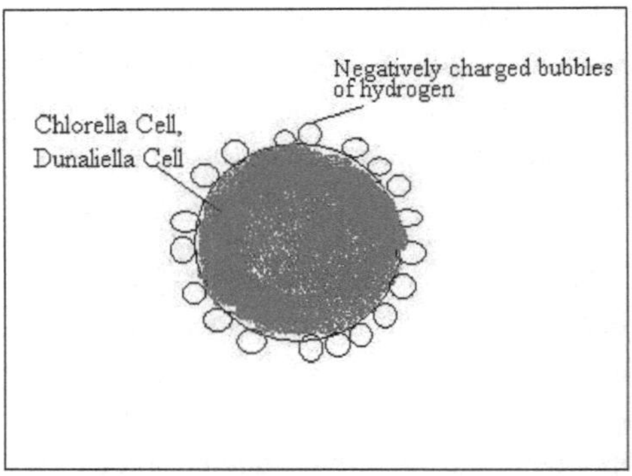

Figure 47: Complete coverage of the entire positively charged lipid surface of the membrane of a Chlorella or Dunaliella cell by negatively charged hydrogen bubbles if the cell size is much larger than the hydrogen bubble size and a negative shield is formed of the complex.

It should be noted that intensive formation of negatively charged microdispersed electrolytic hydrogen bubbles, their repulsion from the cathode and rise to the free surface of microalgae cells and nutrient medium, saturates water with hydrogen gas bubbles, forming negatively charged hydrogen water - catholyte solution (living water).

As established above, the negatively charged hydrogen solution electrostatically enhances positive charges on the outer side of the cell membrane of a cell with negative charges on the inner side, which helps to open the cholesterol contained within the pores of the lipid bilayer, resulting in an enhanced increase in the transport of water, oxygen, carbon dioxide, and lipophilic substances into the biological cell.

The accelerated release of ***Na***+ 1 cations from the biological cell through the transport protein is accomplished by the electrostatic attraction of the negatively charged catholyte to the released positively charged ***Na***+ 1 cations.

Extracellular ***K+1*** cations bind to the protein, causing the release of a phosphate group, the loss of which restores the original conformation of the protein, and releases ***K+1*** cations into the cell.

Therefore, living water, being a catholyte, has enhanced biostimulant properties.

It has a beneficial effect on the Chlorella or Dunaliella cell, activating metabolic processes, increasing its defenses.

Thus, the developed universal electric separator-concentrator uses the effect of direct electric current on the nutrient medium with microalgae, which allows to simultaneously generate microdispersed negative hydrogen bubbles and negatively charged catholyte near the cathode, which as a result of electrostatic force of attraction between the electrodes form strong complexes: live microalgae cell + microdispersed hydrogen bubbles, concentrated throughout the anode zone, separated from the microalgae by the electrostatic force of attraction between the electrodes.

The conical shape of the chamber containing the complexes: live microalgae cells + microdispersed hydrogen bubbles, concentrated along the entire anode zone, and nutrient medium localized along the entire cathode zone, allows, as mentioned earlier, to move the separated components into different ampoules in microgravity due to the resulting surface tension force directed downward to the small base of the cone and equal to the difference between the surface tension force applied to the surface of the common mixture of the large and small base of the cone.

The general appearance of the universal electric separator-concentrator operating according to the previously described principles in microgravity is similar to the design of the universal electric separator presented above (Figures 40, 41), shown in Figure 48.

The concentrator works as follows.

The universal electric separator-concentrator is pre-filled with Chlorella or Dunaliella cells and their nutrient medium from the photobiogenerator (9) described above in a closed cycle. At the same time, the gate of the universal electric separator (3) is closed to separate the medium from the receiving capsule-collectors (5) by the control device that has received a signal from the photocell (transparent color) located in the lower part of the grid cathode.

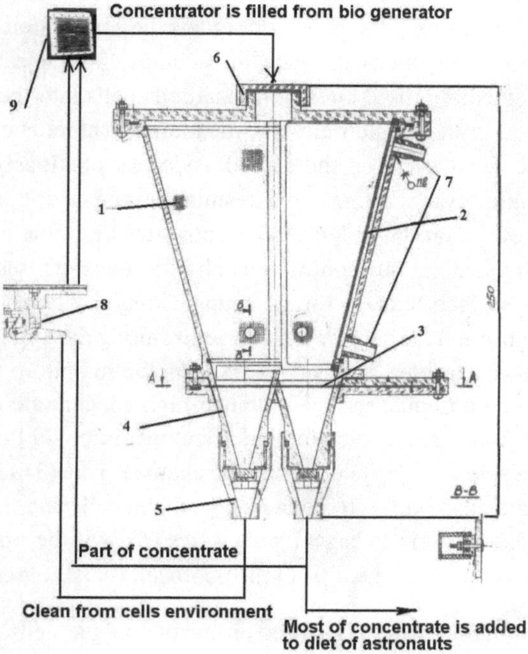

Figure 48: Universal electrical separator-concentrator
1, 2 - metal mesh electrodes; 3 - Gate with louver mechanism;
4 - Sampling outlet taps; 5 - Receiving manifold capsules;
6 - Cover; 7 - Pipes for flushing the separator after the end of work;
8 - Shutter release mechanism; 9 - Biogenerator.

When an electric voltage is applied to the electrodes of the metal grid (1, 2) of the chamber containing a mixture of Chlorella or Dunaliella cells or Dunaliella cells and their nutrient medium, in the range of 26-36 V, the process of electrolysis of Chlorella or Dunaliella cells or Dunaliella cells and their nutrient medium and the formation of microdispersed hydrogen bubbles takes place.

Microdispersed negatively charged hydrogen bubbles meet a Chlorella or Dunaliella cell on their way, adhere to it, forming a negative screen on it, as a result of which the formed complexes of microdispersed negatively charged hydrogen bubbles + Chlorella or Dunaliella cell move towards the reticulated anode due to its positive charge.

At the moment of displacement of the largest number of complexes on the green anode (separation), the photocell signals the control unit (dark color) and the control unit opens the gate to withdraw and select the highly concentrated medium of Chlorella or Dunaliella cells into the capsule (5) on the anode side and the nutrient-free medium of Chlorella or Dunaliella cells into the capsule (5) on the cathode side. As previously described, under microgravity conditions, the resultant force acting on the high concentrate of Chlorella or Dunaliella cells and the remaining nutrient medium contained in the conical chamber is equal to the difference between the surface tension forces acting along the large base of the conical chamber and the surface tension acting along the large bases of the selected conical capsules, and is directed from top to bottom, as shown in Figure 44, and consequently, the obtained high concentrate of Chlorella or Dunaliella cells and the remaining nutrient medium will flow smoothly into different selective capsules, while the chamber gate (3) separating the contents of the chamber - Chlorella or Dunaliella cells and the remaining nutrient medium from the selective capsules (5), will be opened by the control unit receiving a signal from the photocell (dark color).

All of the nutrient medium purified from microalgae cells and a small portion of the high concentration of live microalgae cells is returned to the photobiogenerator to maintain microalgae cultivation for extended periods of time in a closed loop; Most of the high concentration of live microalgae cells is added to the astronauts' diet during long-duration flight, as shown in Figure 48.

At the moment when the Chlorella or Dunaliella cells and the remaining nutrient medium flow out of the chamber, the photocell gives a signal to the control unit (transparent color) and the control unit closes the shutter (5), allowing the chamber to fill with microalgae cells and their nutrient medium from the photobiogenerator, and the process thus starts a new cycle.
The newly developed life support system (LSS) for the International Space Station (ISS) or interplanetary spacecraft (Figure 49) includes the previously described set of universal

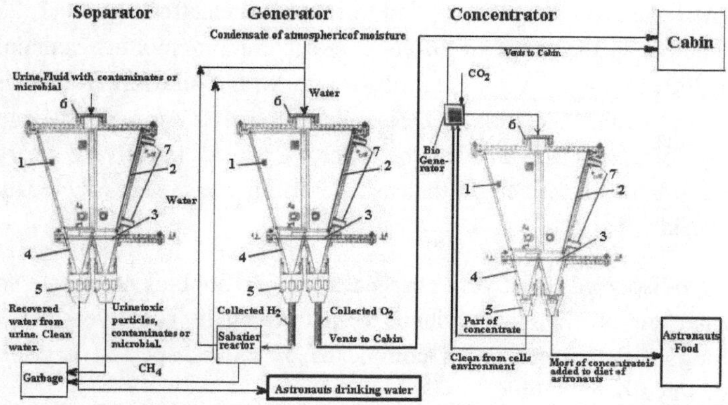

Figure 49: Schematic diagram of the life support system on board an international space station or interplanetary spacecraft.

electric separators, universal electric separators for obtaining clean fuel liquid and drinking water by purifying fuel liquid and technical dirty water from solid harmful impurities or microbes, recovering drinking water from astronauts' urine under weightlessness conditions in quantities sufficient for consumption of this water by astronauts during long space flights; universal electric separator-generator allowing economical and simple separate production of oxygen, hydrogen and drinking water for replenishment of the atmosphere; universal electric separator-generator, allowing economical and simple separate production of oxygen, hydrogen and drinking water for replenishment of the atmosphere; universal electric separator-generator-generator, allowing economical and simple separate production of oxygen, hydrogen and drinking water for replenishment of the atmosphere

The Life Support System (LSS) operates continuously as follows.

1. The universal electric separator is pre-filled with appropriate medium: urine or dirty process water with micro-particles of contaminants or microbial cells. At the same time, the separator gate (3) is closed to separate the total water mixture from the receiving capsule-collectors (5), by feeding a signal from the photocell located in the lower part of the cathode grid (on the left side of the chamber) to the control unit (transparent color).

When electric voltage is applied to the metal electrode grids (1, 2) of the chamber in the range of 26-36 V in a common mixture containing water, the process of water electrolysis and formation of microdispersed hydrogen bubbles takes place. Due to the electrostatic force between the cathode grid and the anode, the microdispersed negatively charged hydrogen bubbles formed on the cathode grid move towards the positive anode grid.

Microdispersed negatively charged hydrogen bubbles meeting a solid microparticle on their way adhere to it, forming its negative shielding, resulting in the formation of complexes of microdispersed negatively charged hydrogen bubbles

hydrogen bubbles + microparticle move towards the reticulated anode due to its positive charge.

The conical shape of the chamber containing the separated complexes: microdispersed negatively charged hydrogen bubbles + microparticle and drinking water, allows to move these components into different ampoules in microgravity conditions due to the action of the resulting surface tension force directed downward to the small base of the cone and equal to difference between the surface tension force applied to the surface of the common mixture of the large base of the cone and the surface tension force,

applied to the surface of the total mixture of the small base of the cone.

At the moment of accumulation of the largest amount of complexes on the grid anode (separation) and clarification on the grid cathode, the photocell gives a signal to the control unit (dark color on the grid anode), which opens the gate for withdrawal and collection of toxic urine microparticles or solid contaminants and microbes into the capsule (5) on the anode side, which go into a waste tank, and clean drinking water, free from solid microparticles, into the capsule (5) on the cathode side, which partially fills the tank of drinking water used for astronauts' drinking, and the other part of it goes into the generator to produce gaseous oxygen and gaseous hydrogen.

2. From above, the universal electric separator-generator is pre-filled with reconstituted drinking water from urine or drinking water free of solid

micro-particles and microbes, or atmospheric moisture condensate. At the same time, the gate (3) of the universal electric separator-generator separating the contents of the chamber water from the capsules (5) is closed due to the signal from the photocell located in the lower part of the cathode grid (on the left side of the chamber) sent to the control unit (transparent color).

When an electric voltage in the range of 26-36 V is applied to the metal grids of the electrodes (1, 2), water electrolysis occurs and microdispersed electrolytic negatively charged hydrogen bubbles and positively charged oxygen bubbles are formed.

Due to the electrostatic attraction force acting between the cathode grid and the positively charged oxygen bubbles formed at the anode, they move horizontally to the near-cathode zone across the cathode surface.

Under microgravity conditions, the resulting high concentration of Chlorella or Dunaliella cells and the remaining nutrient medium will flow smoothly into the various selective capsules, while the chamber shutter (3) separating the chamber contents - Chlorella or Dunaliella cells and the remaining nutrient medium from the selective capsules (5), will be opened by the control unit receiving a signal from the photocell (dark color).

All of the nutrient medium purified from microalgae cells and a small portion of the high concentration of live microalgae cells is returned to the photobiogenerator to sustain microalgae cultivation for long periods of time in a closed loop; most of the high concentration of live microalgae cells is added to the diet of astronauts during long-duration flight.

At the moment when the Chlorella or Dunaliella cells and the remaining nutrient medium flow out of the chamber, the photocell gives a signal to the control unit (transparent color) and the control unit closes the shutter (5), allowing the chamber to fill with microalgae cells and their nutrient medium from the photobiogenerator, and the process thus starts a new cycle.

Thus, the newly developed life support system (LSS) of the International Space Station (ISS) or an interplanetary spacecraft makes it possible to continuously enrich the cosmonauts' atmosphere with oxygen using the system generator, the source of which is drinking water obtained by the system separator by purification of the cosmonauts' vital fluids: recovery of drinking water from urine, purification of dirty technical water used by the cosmonauts from contaminants and microbes, obtaining drinking water by condensation of atmospheric water, purification of drinking water by condensation of atmospheric water, purification of water used by the cosmonauts from contaminants and microbes, and purification of drinking water by condensation of atmospheric water.

The concentrator of the system allows to obtain a concentrate of cells of rapidly proliferating microalgae of high quality, which is used as an energy supplement in the diet of astronauts, from a mixture of microalgae cells and their nutrient medium obtained from a photobioreactor that utilizes carbon dioxide and generates oxygen by photosynthesis of microalgae biomass.

Recirculating a portion of the rapidly proliferating microalgae cell concentrate and cell-purified nutrient medium back into the photobioreactor allows microalgae cells to be grown for a long enough time for extended spaceflight.

Tests of individual components of the life support system (LSS): separator, generator, and concentrator were conducted aboard the IL-76K flying laboratory under short-term weightlessness conditions [21].

The tests confirmed the performance of individual components of the life support system (LSS), and consequently, the system itself in short-term weightlessness.

The frames (Figure 50) show the operation of the universal electric separator under short-term weightlessness conditions.

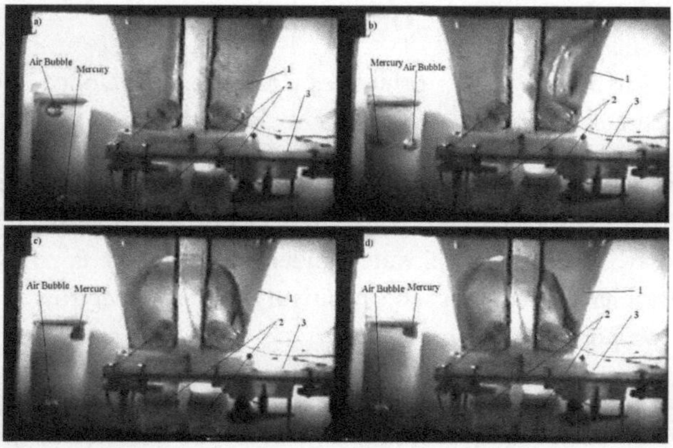

Figure 50 a, b, c, d: Film footage captures the operation of the separator under short-term weightlessness conditions.
a, b, c and d - microgravity in time;
1- a chamber with two metal mesh electrodes;
2- Receiving collector capsules;
3- shutter with louver mechanism.

Thus, a new life support system (LSS) for the International Space Station (ISS) or interplanetary spacecraft has been developed, incorporating previously developed: A universal electric separator for purifying fuel fluid and obtaining drinking water by purifying fuel fluid and dirty technical water used by astronauts from solid harmful impurities or microbes, as well as obtaining drinking water from astronauts' urine under weightlessness conditions in quantities sufficient for consumption of this water by astronauts in long space flights; A universal electric separator-generator that allows economical and simple separate production of oxygen, hydrogen and drinking water for replenishment of the

The developed life support system (LSS) for the International Space Station (ISS) or an interplanetary spacecraft has successfully passed flight tests on board the IL-76K flying laboratory and demonstrated its effective operation in short-term weightlessness.

Round-the-clock functioning of the developed life support system (LSS) of the International Space Station is provided by continuous supply of small voltage to the electrodes of the complex of universal separators from solar batteries of the International Space Station and continuous supply of appropriate fluids to the universal electric separators.

CONCLUSION

1- The main physical principles of the developed universal electroseparator operating in microgravity conditions are:

- Theoretically described and experimentally confirmed negatively charged microdispersed microdispersed electrolytic hydrogen bubbles produced by electrolysis of water under microgravity conditions at a negatively charged cathode and positively charged electrolytic oxygen bubbles at a positively charged anode;

- Charges on the surface of bubbles allow them to move horizontally in water under microgravity conditions - hydrogen bubbles to the anode, oxygen bubbles to the cathode due to the action of electrostatic force of attraction;

- Negatively charged microdispersed hydrogen bubbles, meeting on their way dirty microparticles, toxic urine microparticles, microbial cells and microalgae cells, induce positive charges on the surface of microparticles and, as a result of the resulting electrostatic force of attraction and the surface tension force acting in the same direction, are fixed on them, forming, due to the multiple difference in the size of the microparticle and microdispersed hydrogen bubbles, strong complexes: hydrogen bubbles + microparticle; in the case of microbial cells or microalgae cells, negatively charged microdispersed hydrogen bubbles are fixed on the positively charged outer layer of the phospholipid bilayer of the microbial cell or microalgae membrane membrane using the action of electrostatic attraction force and surface tension force, forming strong complexes: hydrogen bubbles + microbial cell or microalgae cell;

- The formed complexes due to negative shielding by adhesion of microbes or microalgae cells to a micro-particle or cell of the whole complex of microdispersed negatively charged hydrogen bubbles acquire negative total charge and due to electrostatic force of attraction of the positive anode move to the anode and localize in its zone (separation), oxygen bubbles localize in the cathode zone.

- The conical shape of the universal electroseparator and its collectors allows, due to the difference in favor of the surface tension force acting on the surface of the liquid with the large base of the cone directed downwards, to move the separated liquid in microgravity conditions into separate collection capsules: on the left side - a capsule with clean water and oxygen bubbles, on the right side - a capsule with a foam of concentrated dirty microparticles, toxic urine microparticles, with a foam of highly concentrated microbial cells and microalgae.

2. tests conducted under short-term weightlessness conditions showed the reliability of the developed universal electroseparator and the possibility of realizing the process of water recovery from urine in the life support system of the International Space Station.

Thus, the developed universal electroseparator, being simple in operation and design, is economically efficient and productive, is operable in space flight conditions and allows practical realization of water recovery from urine in the life support system of the International Space Station.

Round-the-clock operation of the universal electroseparator is ensured by continuous supply of a small voltage to the electrodes of the universal electroseparator from the solar panels of the International Space Station and continuous supply of urine to the universal electroseparator.

3. The developed universal electroseparator using in its design the effect of direct electric current on liquid fuel, technical water and a special form of construction. allows to clean fuel liquid from contaminants or microbes in the tank with liquid fuel of the ISS (International Space Station) power supply system or special fluids of the ISS life support system for a long time in microgravity conditions simply and economically.

Round-the-clock operation of the universal separator is ensured by continuous supply of a small voltage to the electrodes of the universal electric separator from the solar panels of the International Space Station and continuous supply of dirty liquid fuel or dirty technical fluid to the universal electric separator.

4. The conducted tests showed the reliability of the concentrator operation in microgravity conditions and the possibility of its practical use to replenish the photobiogenerator with the concentrate of rapidly multiplying microalgae cells and its nutrient medium in order to grow enough biomass for a long flight, which allows to maintain a normal

atmosphere on the ISS by utilizing carbon dioxide and generating oxygen in it.

Concentrate of rapidly multiplying microalgae cells, obtained by the concentrator in sufficient quantity, used as a high-calorie supplement in the astronauts' diet, allows to maintain the astronauts' performance during long space expeditions. Round-the-clock operation

REFERENCES

1. Romanov S.Yu., Guzenberg A.S. CosmiICal Life Support Systems: Basic life
support developmentrequirements crews of long-term space expeditions // Engineering ecology, 2013; 2(110): 2-15.

2. TsNIIMash (2000) Photo bioreactor. [Online] Available from https://tsniimash.ru/science/scientific-experiments-onboard-the-iss/cnts/experiments/fotobioreaktor/

3.PucknatA.Water recovery study-electrolytic purification of urine contractor
report, 10,1964

4. Brown D, Lindstrom R, Smith J. The Recovery of Water from Urine by MembraneElectrodialysis, Ionics Inc. Cambridge MA, 1963,

5. Putnam D., Russell S., Birbara Ph. Recovery of water from urine by chromium iontreatment and distillation, Patent US3556949A, 1964.

6. Budininkas P. Four-man rated dual catalyst system for the
recovery ofwater from urine, [Final Report], 1978.

7. Samsonov N.M., Bobe L.S., V.M., Farafonov N.S.,·PinskyB.Ja.,Rakov V.V., Margulis V.I., Rifert V.G., GrigorievJu. I., Komolov V.V., Protasov
N.N. Developmentand testing of a vacuum distillation subsystem for water
reclamation from urine, SAETechnical Paper, 1999.

8. Holder D.W. and Hutchens C.F.. Development Status of the International
Space StationUrine Processor Assembly, NASA, Marshall Space Flight Center, 2003.

9. Roy Langton, Chuck Clark, Martin Hewitt, Lonnie Richards, Aircraft fuels
systems, Encyclopedia of Aerospace Engineering, 2010.

10. Booth F, Bruce R. Centrifugal lyophobic separator Patent, US3828524A,
1972.

11.Ted Hill, Microbial growth in aviation fuels, *Aircraft Engineering and Aerospace Technology, 2003.*

12. Christopher J McNamara, Thomas D Perry IV, Ryan Leard, KtistenBearce,

James Dante, Ralph Mitchell, Corrosion of aluminum alloy 2024 by microorganisms isolated from aircraft fuel tanks,Biofouling, 2005, 21 (5-6), 257-265.
13. Sitnikov N. Yu.Control of autotrophic biosynthesis during cultivation of
microalga (Spirulinaplatensis).Series "Technology of storage and processing of agricultural products", 2011, 4, 1.
14. TsNIIMash (2000) Photobioreactor. [Online] Available from https://tsniimash.ru/science/scientific-experiments-onboard-the-isrs/cnts/experiments/fotobioreactor/
15. Petrovichev M.A., Gurtov A.S. Power supply system of the spacecraft onboard complex,Samara, 2007, Available at: *http://repo.ssau.ru/handle/Uchebnye-posobiya/Sistema-energosnabzheniya-bortovogo-kompleksa-kosmicheskih-apparatov-Elektronnyi-resurs-ucheb-posobie-54577*
16. Gavrilov D.S. Hydrogen fuel cells, Semiconductor physics. Microelectronics.Radioelectronics Devices, 2019; 47-50., Available at: *https://elibrary.ru/item.asp?id=42308056*
17. Khudyakov S.A.Development of power plants based on alkaline fuel cells
for the lunarorbital vehicle and the Buran reusable spacecraft, Alternative Energy and Ecology, 2002; 37-62., Available at: *https://cyberleninka.ru/article/n/razrabotka-energoustanovok-na-osnove-schelochnyh-toplivnyh-elementov-dlya-lunnogo-orbitalnogo-korablya-i-mnogorazovogo*
18. Lessons from Space Accidents: Defeat and Triumph of Apollo 13, 20.04.
2014. Available at:https://habr.com/ru/post/220049/
19. Shoikhedbrod M. Mechanism of Formation of Electrolysis Bubbles in Water
and the Use of their Properties for the Development of New Technologies in
Microgravity, Journal of Fluid Mechanics and Mechanical Design. 2021; 3(3).
20. Frumkin A.N.The selected transactions: electrode processes. M.,*Science*,
1987, Available at: https://www.twirpx.com/file/1621713/
http://techjournals.stmjournals.in/index.php/JoAET/issue/view/212.

21,Shoikhedbrod M.P. Behavior of water under the influence of vibration and
electric field,Lambert Academic Publishing, Toronto, 2017
22. Shoikhedbrod M.P. The Impact of a Negatively or Positively Charged Liquid on the Membrane of a Biological Cell Placed in It, International Journal of Cell Biology and Cellular Processes, 2021; 7(2): 18-29.
23. Proshkin V.Yu., Kurmazenko E.A. Ocygen Generation System "Electron -
VM" on Board ofthe International Space Station, Manned Flights to Space,
2013, 3(8), 84, Available at: https://echemistry.ru/literatura/tezisy/elektroliz-vody-dlya-polucheniya-kisloroda-v-sisteme-elektron-vm-na-mezhdunarodnoj-kosmicheskoj-stancii.html
24. Erickson R., Howe J., Kulp G., Keuren S. International Space Station United States OrbitalSegment Oxygen Generation Systems On-orbit Operation Experience // 38th InternationalConference on Environmental Systems (San Francisco, CA, USA), June 29-July 2, 2008,SAE Publication. No. 2008-01-1962, 1-10., Available at: https://www.sae.org/publications/technical-papers/content/2008-01-1962/
25. Bagdigan R.M., Clode D. Status of the International Space Station Regenerative ECLSSWater Recovery and Oxygen Generation Systems // 35th International Conference onEnvironmental Systems and 8th European
Symposium on Space Environmental ControlSystems (Rome, Italy), July 11-14, 2005, SAE Publication. No. 2005-01-2779, 1-9pp., Available at: https://www.sae.org/publications/technical-papers/content/2005-01-2779/
26. Cloud D., Devin M., Schneider S. et al. ISS Oxygen Generation Design
Status // 29thInternational Conference on Environmental Systems (Denver, Colorado, USA), July 12-15,1999, ICES Publication, No. 1999-01-2116, 1
-10., Available at:
https://www.sae.org/publications/technical-papers/content/1999-01-2116/

27. Shoikhedbrod M.P.New Possibilities for Generating Oxygen and Hydrogen
Using Electrolysisin Microgravity, Journal of Recent Trends in Mechanics,
2021; 6(3): 9-2., Available at: https://matjournals.in/index.php/JoRTM/article/view/7429
28. Shoikhedbrod M.P. The Recovery Water from Urine in the Life-Support
System of automatic Spacecraft using Electric Field Effect on the Urine in The Conditions of Space Flight, Journal of Aerospace Engineering & Technology, ISSN: 2231-038X (Online), ISSN:2348-7887 (Print), 2020, 10(1), 47-53.,Available at: https://techjournals.stmjournals.in/index.php/JoAET/article/view/820

FSC
www.fsc.org
MIX
Papier | Fördert
gute Waldnutzung
FSC® C083411

Kolibri 360 GmbH
Ferdinand-Jühlke-Straße 7
99095 Erfurt, Deutschland
produktsicherheit@kolibri360.de

Druck:
CPI Druckdienstleistungen GmbH
im Auftrag der
Kolibri 360 GmbH
Ferdinand-Jühlke-Str. 7
99095 Erfurt